HOLT SCIENCE & TECHNOLOGY

Environmental Science

HOLT, RINEHART AND WINSTON

A Harcourt Classroom Education Company

Austin · New York · Orlando · Atlanta · San Francisco · Boston · Dallas · Toronto · London

Acknowledgments

Chapter Writers

Katy Z. Allen
Science Writer and Former Biology Teacher
Wayland, Massachusetts

Linda Ruth Berg, Ph.D.
Adjunct Professor–Natural Sciences
St. Petersburg Junior College
St. Petersburg, Florida

Jennie Dusheck
Science Writer
Santa Cruz, California

Mark F. Taylor, Ph.D.
Associate Professor of Biology
Baylor University
Waco, Texas

Lab Writers

Diana Scheidle Bartos
Science Consultant and Educator
Diana Scheidle Bartos, L.L.C.
Lakewood, Colorado

Carl Benson
General Science Teacher
Plains High School
Plains, Montana

Charlotte Blassingame
Technology Coordinator
White Station Middle School
Memphis, Tennessee

Marsha Carver
Science Teacher and Dept. Chair
McLean County High School
Calhoun, Kentucky

Kenneth E. Creese
Science Teacher
White Mountain Junior High School
Rock Springs, Wyoming

Linda Culp
Science Teacher and Dept. Chair
Thorndale High School
Thorndale, Texas

James Deaver
Science Teacher and Dept. Chair
West Point High School
West Point, Nebraska

Frank McKinney, Ph.D.
Professor of Geology
Appalachian State University
Boone, North Carolina

Alyson Mike
Science Teacher
East Valley Middle School
East Helena, Montana

C. Ford Morishita
Biology Teacher
Clackamas High School
Milwaukie, Oregon

Patricia D. Morrell, Ph.D.
Assistant Professor, School of Education
University of Portland
Portland, Oregon

Hilary C. Olson, Ph.D.
Research Associate
Institute for Geophysics
The University of Texas
Austin, Texas

James B. Pulley
Science Editor and Former Science Teacher
Liberty High School
Liberty, Missouri

Denice Lee Sandefur
Science Chairperson
Nucla High School
Nucla, Colorado

Patti Soderberg
Science Writer
The BioQUEST Curriculum Consortium
Beloit College
Beloit, Wisconsin

Phillip Vavala
Science Teacher and Dept. Chair
Salesianum School
Wilmington, Delaware

Albert C. Wartski
Biology Teacher
Chapel Hill High School
Chapel Hill, North Carolina

Lynn Marie Wartski
Science Writer and Former Science Teacher
Hillsborough, North Carolina

Ivora D. Washington
Science Teacher and Dept. Chair
Hyattsville Middle School
Washington, D.C.

Academic Reviewers

Renato J. Aguilera, Ph.D.
Associate Professor
Department of Molecular, Cell, and Developmental Biology
University of California
Los Angeles, California

David M. Armstrong, Ph.D.
Professor of Biology
Department of E.P.O. Biology
University of Colorado
Boulder, Colorado

Alissa Arp, Ph.D.
Director and Professor of Environmental Studies
Romberg Tiburon Center
San Francisco State University
Tiburon, California

Russell M. Brengelman
Professor of Physics
Morehead State University
Morehead, Kentucky

John A. Brockhaus, Ph.D.
Director of Mapping, Charting, and Geodesy Program
Department of Geography and Environmental Engineering
United States Military Academy
West Point, New York

Linda K. Butler, Ph.D.
Lecturer of Biological Sciences
The University of Texas
Austin, Texas

Barry Chernoff, Ph.D.
Associate Curator
Division of Fishes
The Field Museum of Natural History
Chicago, Illinois

Donna Greenwood Crenshaw, Ph.D.
Instructor
Department of Biology
Duke University
Durham, North Carolina

Hugh Crenshaw, Ph.D.
Assistant Professor of Zoology
Duke University
Durham, North Carolina

Joe W. Crim, Ph.D.
Professor of Biology
University of Georgia
Athens, Georgia

Peter Demmin, Ed.D.
Former Science Teacher and Chair
Amherst Central High School
Amherst, New York

Joseph L. Graves, Jr., Ph.D.
Associate Professor of Evolutionary Biology
Arizona State University West
Phoenix, Arizona

William B. Guggino, Ph.D.
Professor of Physiology and Pediatrics
The Johns Hopkins University School of Medicine
Baltimore, Maryland

David Haig, Ph.D.
Assistant Professor of Biology
Department of Organismic and Evolutionary Biology
Harvard University
Cambridge, Massachusetts

Roy W. Hann, Jr., Ph.D.
Professor of Civil Engineering
Texas A&M University
College Station, Texas

Acknowledgments (cont.)

John E. Hoover, Ph.D.
Associate Professor of Biology
Millersville University
Millersville, Pennsylvania

Joan E. N. Hudson, Ph.D.
Associate Professor of Biological Sciences
Sam Houston State University
Huntsville, Texas

Laurie Jackson-Grusby, Ph.D.
Research Scientist and Doctoral Associate
Whitehead Institute for Biomedical Research
Massachusetts Institute of Technology
Cambridge, Massachusetts

George M. Langford, Ph.D.
Professor of Biological Sciences
Dartmouth College
Hanover, New Hampshire

Melanie C. Lewis, Ph.D.
Professor of Biology, Retired
Southwest Texas State University
San Marcos, Texas

V. Patteson Lombardi, Ph.D.
Research Assistant Professor of Biology
Department of Biology
University of Oregon
Eugene, Oregon

Glen Longley, Ph.D.
Professor of Biology and Director of the Edwards Aquifer Research Center
Southwest Texas State University
San Marcos, Texas

William F. McComas, Ph.D.
Director of the Center to Advance Science Education
University of Southern California
Los Angeles, California

LaMoine L. Motz, Ph.D.
Coordinator of Science Education
Oakland County Schools
Waterford, Michigan

Nancy Parker, Ph.D.
Associate Professor of Biology
Southern Illinois University
Edwardsville, Illinois

Barron S. Rector, Ph.D.
Associate Professor and Extension Range Specialist
Texas Agricultural Extension Service
Texas A&M University
College Station, Texas

Peter Sheridan, Ph.D.
Professor of Chemistry
Colgate University
Hamilton, New York

Miles R. Silman, Ph.D.
Assistant Professor of Biology
Wake Forest University
Winston-Salem, North Carolina

Neil Simister, Ph.D.
Associate Professor of Biology
Department of Life Sciences
Brandeis University
Waltham, Massachusetts

Lee Smith, Ph.D.
Curriculum Writer
MDL Information Systems, Inc.
San Leandro, California

Robert G. Steen, Ph.D.
Manager, Rat Genome Project
Whitehead Institute—Center for Genome Research
Massachusetts Institute of Technology
Cambridge, Massachusetts

Martin VanDyke, Ph.D.
Professor of Chemistry, Emeritus
Front Range Community College
Westminister, Colorado

E. Peter Volpe, Ph.D.
Professor of Medical Genetics
Mercer University School of Medicine
Macon, Georgia

Harold K. Voris, Ph.D.
Curator and Head
Division of Amphibians and Reptiles
The Field Museum of Natural History
Chicago, Illinois

Mollie Walton
Biology Instructor
El Paso Community College
El Paso, Texas

Peter Wetherwax, Ph.D.
Professor of Biology
University of Oregon
Eugene, Oregon

Mary K. Wicksten, Ph.D.
Professor of Biology
Texas A&M University
College Station, Texas

R. Stimson Wilcox, Ph.D.
Associate Professor of Biology
Department of Biological Sciences
Binghamton University
Binghamton, New York

Conrad M. Zapanta, Ph.D.
Research Engineer
Sulzer Carbomedics, Inc.
Austin, Texas

Safety Reviewer

Jack Gerlovich, Ph.D.
Associate Professor
School of Education
Drake University
Des Moines, Iowa

Teacher Reviewers

Barry L. Bishop
Science Teacher and Dept. Chair
San Rafael Junior High School
Ferron, Utah

Carol A. Bornhorst
Science Teacher and Dept. Chair
Bonita Vista Middle School
Chula Vista, California

Paul Boyle
Science Teacher
Perry Heights Middle School
Evansville, Indiana

Yvonne Brannum
Science Teacher and Dept. Chair
Hine Junior High School
Washington, D.C.

Gladys Cherniak
Science Teacher
St. Paul's Episcopal School
Mobile, Alabama

James Chin
Science Teacher
Frank A. Day Middle School
Newtonville, Massachusetts

Kenneth Creese
Science Teacher
White Mountain Junior High School
Rock Springs, Wyoming

Linda A. Culp
Science Teacher and Dept. Chair
Thorndale High School
Thorndale, Texas

Georgiann Delgadillo
Science Teacher
East Valley Continuous Curriculum School
Spokane, Washington

Alonda Droege
Biology Teacher
Evergreen High School
Seattle, Washington

Michael J. DuPré
Curriculum Specialist
Rush Henrietta Junior-Senior High School
Henrietta, New York

Rebecca Ferguson
Science Teacher
North Ridge Middle School
North Richland Hills, Texas

Susan Gorman
Science Teacher
North Ridge Middle School
North Richland Hills, Texas

Gary Habeeb
Science Mentor
Sierra-Plumas Joint Unified School District
Downieville, California

Karma Houston-Hughes
Science Mentor
Kyrene Middle School
Tempe, Arizona

Roberta Jacobowitz
Science Teacher
C. W. Otto Middle School
Lansing, Michigan

Kerry A. Johnson
Science Teacher
Isbell Middle School
Santa Paula, California

M. R. Penny Kisiah
Science Teacher and Dept. Chair
Fairview Middle School
Tallahassee, Florida

Kathy LaRoe
Science Teacher
East Valley Middle School
East Helena, Montana

Jane M. Lemons
Science Teacher
Western Rockingham Middle School
Madison, North Carolina

Scott Mandel, Ph.D.
Director and Educational Consultant
Teachers Helping Teachers
Los Angeles, California

Thomas Manerchia
Former Biology and Life Science Teacher
Archmere Academy
Claymont, Delaware

Maurine O. Marchani
Science Teacher and Dept. Chair
Raymond Park Middle School
Indianapolis, Indiana

Jason P. Marsh
Biology Teacher
Montevideo High School and Montevideo Country School
Montevideo, Minnesota

Edith C. McAlanis
Science Teacher and Dept. Chair
Socorro Middle School
El Paso, Texas

Kevin McCurdy, Ph.D.
Science Teacher
Elmwood Junior High School
Rogers, Arkansas

Kathy McKee
Science Teacher
Hoyt Middle School
Des Moines, Iowa

E Environmental Science

Skills Development

Process Skills

Skills Development *(continued)*

Research and Critical Thinking Skills

Apply

Feature Articles

Health Watch

Eye on the Environment

Weird Science

Across the Sciences

Careers

Scientific Debate

Eureka!

Connections

Environment Connection

Mathematics

Program Scope and Sequence

Selecting the right books for your course is easy. Just review the topics presented in each book to determine the best match to your district curriculum.

	A MICROORGANISMS, FUNGI, AND PLANTS	**B** ANIMALS	
CHAPTER 1	**It's Alive!! Or, Is It?** ❏ Characteristics of living things ❏ Homeostasis ❏ Heredity and DNA ❏ Producers, consumers, and decomposers ❏ Biomolecules	**Animals and Behavior** ❏ Characteristics of animals ❏ Classification of animals ❏ Animal behavior ❏ Hibernation and estivation ❏ The biological clock ❏ Animal communication ❏ Living in groups	
CHAPTER 2	**Bacteria and Viruses** ❏ Binary fission ❏ Characteristics of bacteria ❏ Nitrogen-fixing bacteria ❏ Antibiotics ❏ Pathogenic bacteria ❏ Characteristics of viruses ❏ Lytic cycle	**Invertebrates** ❏ General characteristics of invertebrates ❏ Types of symmetry ❏ Characteristics of sponges, cnidarians, arthropods, and echinoderms ❏ Flatworms versus roundworms ❏ Types of circulatory systems	
CHAPTER 3	**Protists and Fungi** ❏ Characteristics of protists ❏ Types of algae ❏ Types of protozoa ❏ Protist reproduction ❏ Characteristics of fungi and lichens	**Fishes, Amphibians, and Reptiles** ❏ Characteristics of vertebrates ❏ Structure and kinds of fishes ❏ Development of lungs ❏ Structure and kinds of amphibians and reptiles ❏ Function of the amniotic egg	
CHAPTER 4	**Introduction to Plants** ❏ Characteristics of plants and seeds ❏ Reproduction and classification ❏ Angiosperms versus gymnosperms ❏ Monocots versus dicots ❏ Structure and functions of roots, stems, leaves, and flowers	**Birds and Mammals** ❏ Structure and kinds of birds ❏ Types of feathers ❏ Adaptations for flight ❏ Structure and kinds of mammals ❏ Function of the placenta	
CHAPTER 5	**Plant Processes** ❏ Pollination and fertilization ❏ Dormancy ❏ Photosynthesis ❏ Plant tropisms ❏ Seasonal responses of plants		
CHAPTER 6			
CHAPTER 7			

Life Science

C CELLS, HEREDITY, & CLASSIFICATION

Cells: The Basic Units of Life
- ❏ Cells, tissues, and organs
- ❏ Populations, communities, and ecosystems
- ❏ Cell theory
- ❏ Surface-to-volume ratio
- ❏ Prokaryotic versus eukaryotic cells
- ❏ Cell organelles

The Cell in Action
- ❏ Diffusion and osmosis
- ❏ Passive versus active transport
- ❏ Endocytosis versus exocytosis
- ❏ Photosynthesis
- ❏ Cellular respiration and fermentation
- ❏ Cell cycle

Heredity
- ❏ Dominant versus recessive traits
- ❏ Genes and alleles
- ❏ Genotype, phenotype, the Punnett square and probability
- ❏ Meiosis
- ❏ Determination of sex

Genes and Gene Technology
- ❏ Structure of DNA
- ❏ Protein synthesis
- ❏ Mutations
- ❏ Heredity disorders and genetic counseling

The Evolution of Living Things
- ❏ Adaptations and species
- ❏ Evidence for evolution
- ❏ Darwin's work and natural selection
- ❏ Formation of new species

The History of Life on Earth
- ❏ Geologic time scale and extinctions
- ❏ Plate tectonics
- ❏ Human evolution

Classification
- ❏ Levels of classification
- ❏ Cladistic diagrams
- ❏ Dichotomous keys
- ❏ Characteristics of the six kingdoms

D HUMAN BODY SYSTEMS & HEALTH

Body Organization and Structure
- ❏ Homeostasis
- ❏ Types of tissue
- ❏ Organ systems
- ❏ Structure and function of the skeletal system, muscular system, and integumentary system

Circulation and Respiration
- ❏ Structure and function of the cardiovascular system, lymphatic system, and respiratory system
- ❏ Respiratory disorders

The Digestive and Urinary Systems
- ❏ Structure and function of the digestive system
- ❏ Structure and function of the urinary system

Communication and Control
- ❏ Structure and function of the nervous system and endocrine system
- ❏ The senses
- ❏ Structure and function of the eye and ear

Reproduction and Development
- ❏ Asexual versus sexual reproduction
- ❏ Internal versus external fertilization
- ❏ Structure and function of the human male and female reproductive systems
- ❏ Fertilization, placental development, and embryo growth
- ❏ Stages of human life

Body Defenses and Disease
- ❏ Types of diseases
- ❏ Vaccines and immunity
- ❏ Structure and function of the immune system
- ❏ Autoimmune diseases, cancer, and AIDS

Staying Healthy
- ❏ Nutrition and reading food labels
- ❏ Alcohol and drug effects on the body
- ❏ Hygiene, exercise, and first aid

E ENVIRONMENTAL SCIENCE

Interactions of Living Things
- ❏ Biotic versus abiotic parts of the environment
- ❏ Producers, consumers, and decomposers
- ❏ Food chains and food webs
- ❏ Factors limiting population growth
- ❏ Predator-prey relationships
- ❏ Symbiosis and coevolution

Cycles in Nature
- ❏ Water cycle
- ❏ Carbon cycle
- ❏ Nitrogen cycle
- ❏ Ecological succession

The Earth's Ecosystems
- ❏ Kinds of land and water biomes
- ❏ Marine ecosystems
- ❏ Freshwater ecosystems

Environmental Problems and Solutions
- ❏ Types of pollutants
- ❏ Types of resources
- ❏ Conservation practices
- ❏ Species protection

Energy Resources
- ❏ Types of resources
- ❏ Energy resources and pollution
- ❏ Alternative energy resources

Scope and Sequence *(continued)*

	F INSIDE THE RESTLESS EARTH	**G** EARTH'S CHANGING SURFACE
CHAPTER 1	**Minerals of the Earth's Crust** ❏ Mineral composition and structure ❏ Types of minerals ❏ Mineral identification ❏ Mineral formation and mining	**Maps as Models of the Earth** ❏ Structure of a map ❏ Cardinal directions ❏ Latitude, longitude, and the equator ❏ Magnetic declination and true north ❏ Types of projections ❏ Aerial photographs ❏ Remote sensing ❏ Topographic maps
CHAPTER 2	**Rocks: Mineral Mixtures** ❏ Rock cycle and types of rocks ❏ Rock classification ❏ Characteristics of igneous, sedimentary, and metamorphic rocks	**Weathering and Soil Formation** ❏ Types of weathering ❏ Factors affecting the rate of weathering ❏ Composition of soil ❏ Soil conservation and erosion prevention
CHAPTER 3	**The Rock and Fossil Record** ❏ Uniformitarianism versus catastrophism ❏ Superposition ❏ The geologic column and unconformities ❏ Absolute dating and radiometric dating ❏ Characteristics and types of fossils ❏ Geologic time scale	**Agents of Erosion and Deposition** ❏ Shoreline erosion and deposition ❏ Wind erosion and deposition ❏ Erosion and deposition by ice ❏ Gravity's effect on erosion and deposition
CHAPTER 4	**Plate Tectonics** ❏ Structure of the Earth ❏ Continental drifts and sea floor spreading ❏ Plate tectonics theory ❏ Types of boundaries ❏ Types of crust deformities	
CHAPTER 5	**Earthquakes** ❏ Seismology ❏ Features of earthquakes ❏ P and S waves ❏ Gap hypothesis ❏ Earthquake safety	
CHAPTER 6	**Volcanoes** ❏ Types of volcanoes and eruptions ❏ Types of lava and pyroclastic material ❏ Craters versus calderas ❏ Sites and conditions for volcano formation ❏ Predicting eruptions	

Earth Science

H WATER ON EARTH

The Flow of Fresh Water
❏ Water cycle
❏ River systems
❏ Stream erosion
❏ Life cycle of rivers
❏ Deposition
❏ Aquifers, springs, and wells
❏ Ground water
❏ Water treatment and pollution

Exploring the Oceans
❏ Properties and characteristics of the oceans
❏ Features of the ocean floor
❏ Ocean ecology
❏ Ocean resources and pollution

The Movement of Ocean Water
❏ Types of currents
❏ Characteristics of waves
❏ Types of ocean waves
❏ Tides

I WEATHER AND CLIMATE

The Atmosphere
❏ Structure of the atmosphere
❏ Air pressure
❏ Radiation, convection, and conduction
❏ Greenhouse effect and global warming
❏ Characteristics of winds
❏ Types of winds
❏ Air pollution

Understanding Weather
❏ Water cycle
❏ Humidity
❏ Types of clouds
❏ Types of precipitation
❏ Air masses and fronts
❏ Storms, tornadoes, and hurricanes
❏ Weather forecasting
❏ Weather maps

Climate
❏ Weather versus climate
❏ Seasons and latitude
❏ Prevailing winds
❏ Earth's biomes
❏ Earth's climate zones
❏ Ice ages
❏ Global warming
❏ Greenhouse effect

J ASTRONOMY

Observing the Sky
❏ Astronomy
❏ Keeping time
❏ Mapping the stars
❏ Scales of the universe
❏ Types of telescope
❏ Radioastronomy

Formation of the Solar System
❏ Birth of the solar system
❏ Planetary motion
❏ Newton's Law of Universal Gravitation
❏ Structure of the sun
❏ Fusion
❏ Earth's structure and atmosphere

A Family of Planets
❏ Properties and characteristics of the planets
❏ Properties and characteristics of moons
❏ Comets, asteroids, and meteoroids

The Universe Beyond
❏ Composition of stars
❏ Classification of stars
❏ Star brightness, distance, and motions
❏ H-R diagram
❏ Life cycle of stars
❏ Types of galaxies
❏ Theories on the formation of the universe

Exploring Space
❏ Rocketry and artificial satellites
❏ Types of Earth orbit
❏ Space probes and space exploration

Scope and Sequence (continued)

	K INTRODUCTION TO MATTER	**L** INTERACTIONS OF MATTER
CHAPTER 1	**The Properties of Matter** ❏ Definition of matter ❏ Mass and weight ❏ Physical and chemical properties ❏ Physical and chemical change ❏ Density	**Chemical Bonding** ❏ Types of chemical bonds ❏ Valence electrons ❏ Ions versus molecules ❏ Crystal lattice
CHAPTER 2	**States of Matter** ❏ States of matter and their properties ❏ Boyle's and Charles's laws ❏ Changes of state	**Chemical Reactions** ❏ Writing chemical formulas and equations ❏ Law of conservation of mass ❏ Types of reactions ❏ Endothermic versus exothermic reactions ❏ Law of conservation of energy ❏ Activation energy ❏ Catalysts and inhibitors
CHAPTER 3	**Elements, Compounds, and Mixtures** ❏ Elements and compounds ❏ Metals, nonmetals, and metalloids (semiconductors) ❏ Properties of mixtures ❏ Properties of solutions, suspensions, and colloids	**Chemical Compounds** ❏ Ionic versus covalent compounds ❏ Acids, bases, and salts ❏ pH ❏ Organic compounds ❏ Biomolecules
CHAPTER 4	**Introduction to Atoms** ❏ Atomic theory ❏ Atomic model and structure ❏ Isotopes ❏ Atomic mass and mass number	**Atomic Energy** ❏ Properties of radioactive substances ❏ Types of decay ❏ Half-life ❏ Fission, fusion, and chain reactions
CHAPTER 5	**The Periodic Table** ❏ Structure of the periodic table ❏ Periodic law ❏ Properties of alkali metals, alkaline-earth metals, halogens, and noble gases	
CHAPTER 6		

Physical Science

M FORCES, MOTION, AND ENERGY	N ELECTRICITY AND MAGNETISM	O SOUND AND LIGHT

Matter in Motion
❏ Speed, velocity, and acceleration
❏ Measuring force
❏ Friction
❏ Mass versus weight

Introduction to Electricity
❏ Law of electric charges
❏ Conduction versus induction
❏ Static electricity
❏ Potential difference
❏ Cells, batteries, and photocells
❏ Thermocouples
❏ Voltage, current, and resistance
❏ Electric power
❏ Types of circuits

The Energy of Waves
❏ Properties of waves
❏ Types of waves
❏ Reflection and refraction
❏ Diffraction and interference
❏ Standing waves and resonance

Forces in Motion
❏ Terminal velocity and free fall
❏ Projectile motion
❏ Inertia
❏ Momentum

Electromagnetism
❏ Properties of magnets
❏ Magnetic force
❏ Electromagnetism
❏ Solenoids and electric motors
❏ Electromagnetic induction
❏ Generators and transformers

The Nature of Sound
❏ Properties of sound waves
❏ Structure of the human ear
❏ Pitch and the Doppler effect
❏ Infrasonic versus ultrasonic sound
❏ Sound reflection and echolocation
❏ Sound barrier
❏ Interference, resonance, diffraction, and standing waves
❏ Sound quality of instruments

Forces in Fluids
❏ Properties in fluids
❏ Atmospheric pressure
❏ Density
❏ Pascal's principle
❏ Buoyant force
❏ Archimedes' principle
❏ Bernoulli's principle

Electronic Technology
❏ Properties of semiconductors
❏ Integrated circuits
❏ Diodes and transistors
❏ Analog versus digital signals
❏ Microprocessors
❏ Features of computers

The Nature of Light
❏ Electromagnetic waves
❏ Electromagnetic spectrum
❏ Law of reflection
❏ Absorption and scattering
❏ Reflection and refraction
❏ Diffraction and interference

Work and Machines
❏ Measuring work
❏ Measuring power
❏ Types of machines
❏ Mechanical advantage
❏ Mechanical efficiency

Light and Our World
❏ Luminosity
❏ Types of lighting
❏ Types of mirrors and lenses
❏ Focal point
❏ Structure of the human eye
❏ Lasers and holograms

Energy and Energy Resources
❏ Forms of energy
❏ Energy conversions
❏ Law of conservation of energy
❏ Energy resources

Heat and Heat Technology
❏ Heat versus temperature
❏ Thermal expansion
❏ Absolute zero
❏ Conduction, convection, radiation
❏ Conductors versus insulators
❏ Specific heat capacity
❏ Changes of state
❏ Heat engines
❏ Thermal pollution

HOLT SCIENCE & TECHNOLOGY
Components Listing

Effective planning starts with all the resources you need in an easy-to-use package for each short course.

Directed Reading Worksheets Help students develop and practice fundamental reading comprehension skills and provide a comprehensive review tool for students to use when studying for an exam.

Study Guide Vocabulary & Notes Worksheets and Chapter Review Worksheets are reproductions of the Chapter Highlights and Chapter Review sections that follow each chapter in the textbook.

Science Puzzlers, Twisters & Teasers Use vocabulary and concepts from each chapter of the Pupil's Editions as elements of rebuses, anagrams, logic puzzles, daffy definitions, riddle poems, word jumbles, and other types of puzzles.

Reinforcement and Vocabulary Review Worksheets Approach a chapter topic from a different angle with an emphasis on different learning modalities to help students that are frustrated by traditional methods.

Critical Thinking & Problem Solving Worksheets Develop the following skills: distinguishing fact from opinion, predicting consequences, analyzing information, and drawing conclusions. Problem Solving Worksheets develop a step-by-step process of problem analysis including gathering information, asking critical questions, identifying alternatives, and making comparisons.

Math Skills for Science Worksheets Each activity gives a brief introduction to a relevant math skill, a step-by-step explanation of the math process, one or more example problems, and a variety of practice problems.

Science Skills Worksheets Help your students focus specifically on skills such as measuring, graphing, using logic, understanding statistics, organizing research papers, and critical thinking options.

LAB ACTIVITIES

ALL LABS ARE CLASSROOM TESTED & APPROVED

Datasheets for Labs These worksheets are the labs found in the *Holt Science & Technology* textbook. Charts, tables, and graphs are included to make data collection and analysis easier, and space is provided to write observations and conclusions.

Whiz-Bang Demonstrations Discovery or Making Models experiences label each demo as one in which students discover an answer or use a scientific model.

Calculator-Based Labs Give students the opportunity to use graphing-calculator probes and sensors to collect data using a TI graphing calculator, Vernier sensors, and a TI CBL 2™ or Vernier Lab Pro interface.

EcoLabs and Field Activities Focus on educational outdoor projects, such as wildlife observation, nature surveys, or natural history.

Inquiry Labs Use the scientific method to help students find their own path in solving a real-world problem.

Long-Term Projects and Research Ideas Provide students with the opportunity to go beyond library and Internet resources to explore science topics.

ASSESSMENT

Chapter Tests Each four-page chapter test consists of a variety of item types including Multiple Choice, Using Vocabulary, Short Answer, Critical Thinking, Math in Science, Interpreting Graphics, and Concept Mapping.

Performance-Based Assessments Evaluate students' abilities to solve problems using the tools, equipment, and techniques of science. Rubrics included for each assessment make it easy to evaluate student performance.

TEACHER RESOURCES

Lesson Plans Integrate all of the great resources in the *Holt Science & Technology* program into your daily teaching. Each lesson plan includes a correlation of the lesson activities to the National Science Education Standards.

Teaching Transparencies Each transparency is correlated to a particular lesson in the Chapter Organizer.

 **Concept Mapping Transparencies, Worksheets, and Answer Key**

Give students an opportunity to complete their own concept maps to study the concepts within each chapter and form logical connections. Student worksheets contain a blank concept map with linking phrases and a list of terms to be used by the student to complete the map.

TECHNOLOGY RESOURCES

One-Stop Planner CD-ROM

Finding the right resources is easy with the One-Stop Planner CD-ROM. You can view and print any resource with just the click of a mouse. Customize the suggested lesson plans to match your daily or weekly calendar and your district's requirements. Powerful test generator software allows you to create customized assessments using a databank of items.

The One-Stop Planner for each level includes the following:

- All materials from the Teaching Resources
- Bellringer Transparency Masters
- Block Scheduling Tools
- Standards Correlations
- Lab Inventory Checklist
- Safety Information
- Science Fair Guide
- Parent Involvement Tools
- Spanish Audio Scripts
- Spanish Glossary
- Assessment Item Listing
- Assessment Checklists and Rubrics
- Test Generator

sciLINKS

*sci*LINKS numbers throughout the text take you and your students to some of the best on-line resources available. Sites are constantly reviewed and updated by the National Science Teachers Association. Special "teacher only" sites are available to you once you register with the service.

go.hrw.com

To access Holt, Rinehart and Winston Web resources, use the home page codes for each level found on page 1 of the Pupil's Editions. The codes shown on the Chapter Organizers for each chapter in the Annotated Teacher's Edition take you to chapter-specific resources.

Smithsonian Institution

Find lesson plans, activities, interviews, virtual exhibits, and just general information on a wide variety of topics relevant to middle school science.

CNNfyi.com

Find the latest in late-breaking science news for students. Featured news stories are supported with lesson plans and activities.

CNN Presents Science in the News Video Library

Bring relevant science news stories into the classroom. Each video comes with a Teacher's Guide and set of Critical Thinking Worksheets that develop listening and media analysis skills. Tapes in the series include:

- Eye on the Environment
- Multicultural Connections
- Scientists in Action
- Science, Technology & Society

Guided Reading Audio CD Program

Students can listen to a direct read of each chapter and follow along in the text. Use the program as a content bridge for struggling readers and students for whom English is not their native language.

Interactive Explorations CD-ROM

Turn a computer into a virtual laboratory. Students act as lab assistants helping Dr. Crystal Labcoat solve real-world problems. Activities develop students' inquiry, analysis, and decision-making skills.

Interactive Science Encyclopedia CD-ROM

Give your students access to more than 3,000 cross-referenced scientific definitions, in-depth articles, science fair project ideas, activities, and more.

ADDITIONAL COMPONENTS

Holt Anthology of Science Fiction

Science Fiction features in the Pupil's Edition preview the stories found in the anthology. Each story begins with a Reading Prep guide and closes with Think About It questions.

Professional Reference for Teachers

Articles written by leading educators help you learn more about the National Science Education Standards, block scheduling, classroom management techniques, and more. A bibliography of professional references is included.

Holt Science Posters

Seven wall posters highlight interesting topics, such as the Physics of Sports, or useful reference material, such as the Scientific Method.

Holt Science Skills Workshop: Reading in the Content Area

Use a variety of in-depth skills exercises to help students learn to read science materials strategically.

Key
These materials are blackline masters.
■ All titles shown in green are found in the *Teaching Resources* booklets for each course.

Science & Math Skills Worksheets

The *Holt Science and Technology* program helps you meet the needs of a wide variety of students, regardless of their skill level. The following pages provide examples of the worksheets available to improve your students' science and math skills, whether they already have a strong science and math background or are weak in these areas. Samples of assessment checklists and rubrics are also provided.

In addition to the skills worksheets represented here, *Holt Science and Technology* provides a variety of worksheets that are correlated directly with each chapter of the program. Representations of these worksheets are found at the beginning of each chapter in this Annotated Teacher's Edition. Specific worksheets related to each chapter are listed in the Chapter Organizer. Worksheets and transparencies are found in the softcover *Teaching Resources* for each course.

Many worksheets are also available on the HRW Web site. The address is **go.hrw.com.**

Science Skills Worksheets: Thinking Skills

BEING FLEXIBLE
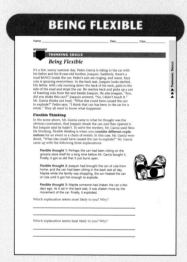

USING YOUR SENSES
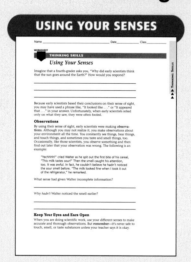

THINKING OBJECTIVELY

UNDERSTANDING BIAS
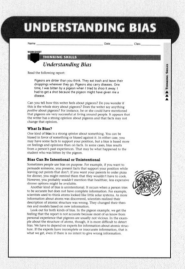

USING LOGIC

BOOSTING YOUR MEMORY
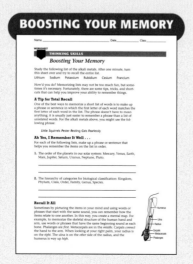

IMPROVING YOUR STUDY HABITS
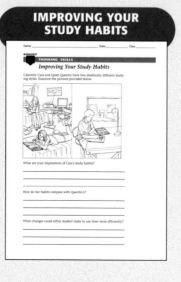

READING A SCIENCE TEXTBOOK
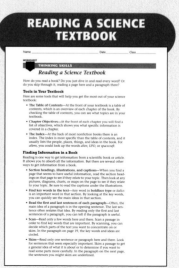

Science Skills Worksheets: Experimenting Skills

SAFETY RULES!

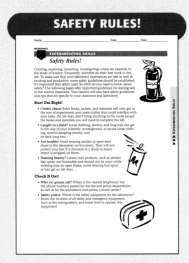

EXPERIMENTING SKILLS
Safety Rules!

DOING A LAB WRITE-UP

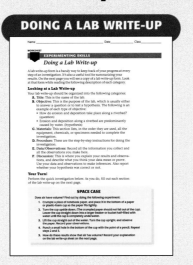

EXPERIMENTING SKILLS
Doing a Lab Write-up

UNDERSTANDING VARIABLES

EXPERIMENTING SKILLS
Understanding Variables

WORKING WITH HYPOTHESES

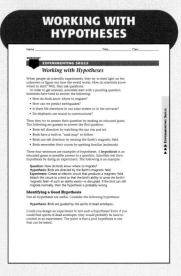

EXPERIMENTING SKILLS
Working with Hypotheses

DESIGNING AN EXPERIMENT

EXPERIMENTING SKILLS
Designing an Experiment

USING THE INTERNATIONAL SYSTEM OF UNITS (SI)

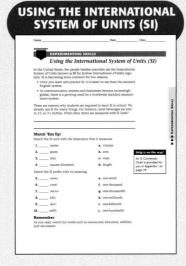

EXPERIMENTING SKILLS
Using the International System of Units (SI)

MEASURING

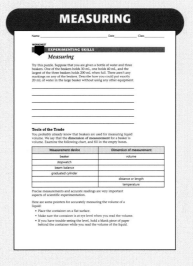

EXPERIMENTING SKILLS
Measuring

Science Skills Worksheets: Researching Skills

CHOOSING YOUR TOPIC

RESEARCHING SKILLS
Choosing Your Topic

ORGANIZING YOUR RESEARCH

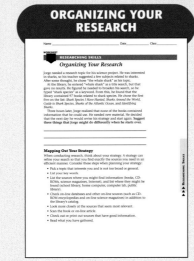

RESEARCHING SKILLS
Organizing Your Research

FINDING USEFUL SOURCES

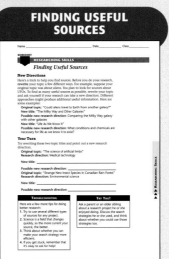

RESEARCHING SKILLS
Finding Useful Sources

RESEARCHING ON THE WEB

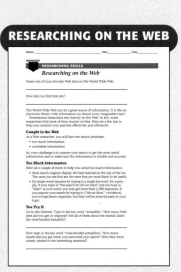

RESEARCHING SKILLS
Researching on the Web

Science & Math Skills Worksheets (continued)

Science Skills Worksheets: Researching Skills (continued)

IDENTIFYING BIAS
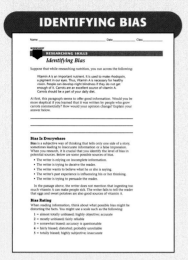

TAKING NOTES
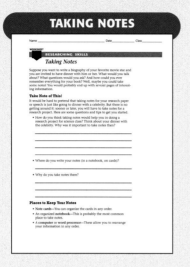

SCIENCE WRITING

Science Skills Worksheets: Communicating Skills

SCIENCE DRAWING
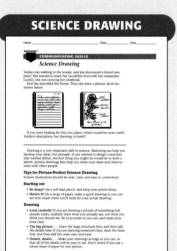

USING MODELS TO COMMUNICATE

INTRODUCTION TO GRAPHS
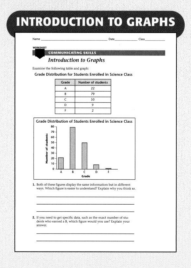

GRASPING GRAPHING
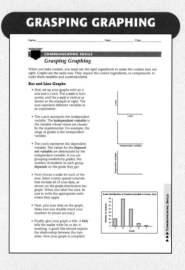

INTERPRETING YOUR DATA
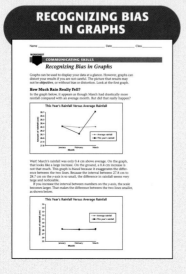

RECOGNIZING BIAS IN GRAPHS
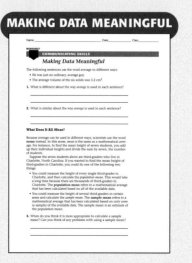

MAKING DATA MEANINGFUL
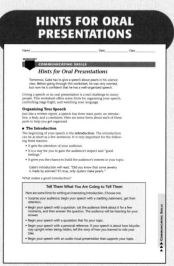

HINTS FOR ORAL PRESENTATIONS

Math Skills for Science

ADDITION AND SUBTRACTION

MATH SKILLS

Addition Review

Addition is used to find the total of two or more quantities. The answer to an addition problem is known as the sum.

PROCEDURE: To find the sum of a set of numbers, align the numbers vertically so that the ones digits are in the same column. Add each column, working from right to left.

SAMPLE PROBLEM: Find the sum of 317, 435, and 92.

Step 1: Add the ones. Don't forget to carry your numbers.

Step 2: Add the tens.

Step 3: Add the hundreds.

The sum is **844.**

Add It Up!

1. Find the sums of the following problems:
 a. 348 + 21   b. 98,125 + 233   c. 593 + 386   d. 36,186 + 27,509

2. Your doctor advises you to take 60 mg of vitamin C, 20 mg of niacin, and 15 mg of zinc every day. How many milligrams of nutrients will you take?

3. A chemistry experiment calls for 356 mL of water, 197 mL of saline solution, and 55 mL of vinegar. How much liquid is needed in all?

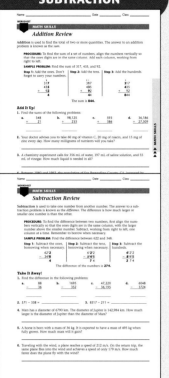

Subtraction Review

Subtraction is used to take one number from another number. The answer to a subtraction problem is known as the difference. The difference is how much larger or smaller one number is than the other.

PROCEDURE: To find the difference between two numbers, first align the numbers vertically so that the ones digits are in the same column, with the larger number above the smaller number. Subtract, working from right to left, one column at a time. Remember to borrow when necessary.

SAMPLE PROBLEM: Find the difference between 622 and 348.

Step 1: Subtract the ones, borrowing when necessary.

Step 2: Subtract the tens, borrowing when necessary.

Step 3: Subtract the hundreds.

The difference of the numbers is **274.**

Take It Away!

1. Find the difference in the following problems:
 a. 88 − 36   b. 1695 − 352   c. 47,220 − 36,195   d. 6048 − 3724

2. 571 − 338 =   3. 8317 − 211 =

4. Mars has a diameter of 6790 km. The diameter of Jupiter is 142,984 km. How much larger is the diameter of Jupiter than the diameter of Mars?

5. A horse is born with a mass of 36 kg. It is expected to have a mass of 495 kg when fully grown. How much mass will it gain?

6. Traveling with the wind, a plane reaches a speed of 212 m/s. On the return trip, the same plane flies into the wind and achieves a speed of only 179 m/s. How much faster does the plane fly with the wind?

MULTIPLICATION

Name _____ Date _____ Class _____

MATH SKILLS

Multiplying Whole Numbers

Suppose every student in your class planted 5 seeds in your school's garden. How many seeds were planted? You could repeatedly add 5 seeds plus 5 seeds until every student's seeds had been added, but this would be pretty time consuming. **Multiplication,** which simplifies addition, is the process of calculating the total of a number that is added together a specific number of times. For example, 3 × 4 means adding 3 together 4 times, or 3 + 3 + 3 + 3 = 12. So 3 × 4 = 12. The answer to a multiplication problem is called the product.

PROCEDURE: To find the product of two whole numbers, align your numbers so that the ones digit is in the same column. Multiply each digit of the top number by the ones digit in the bottom number, carrying when necessary. Then multiply each digit in the top number by the tens in the bottom number, regrouping when necessary. Finally, add the partial products to find the final product.

SAMPLE PROBLEM: Find the product of 34 and 16.

Step 1: Align the numbers vertically. Multiply each digit in the top number by the ones digit in the bottom number. Carry when necessary.

Step 2: Multiply each digit in the top number by the tens in the bottom number. Imagine adding a zero in the ones column as a place holder.

Step 3: Add the partial products.

The product is **544.**

Practice Your Skills!

1. Multiply. Don't forget to show all your work.
 a. 12 × 24   b. 245 × 36   c. 46 × 87   d. 2751 × 11

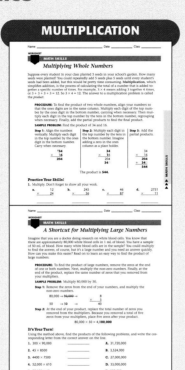

A Shortcut for Multiplying Large Numbers

Imagine that you are a doctor doing research on white blood cells. You know that there are approximately 80,000 white blood cells in 1 mL of blood. You have a sample of 50 mL of blood. How many white blood cells are in the sample? You could multiply to find the answer, of course, but it's a large number and you need an answer quickly. How can you make this easier? Read on to learn an easy way to find the product of large numbers.

PROCEDURE: To find the product of large numbers, remove the zeros at the end of one or both numbers. Next, multiply the non-zero numbers. Finally, at the end of the product, replace the same number of zeros that you removed from your multipliers.

SAMPLE PROBLEM: Multiply 80,000 by 50.

Step 1: Remove the zeros from the end of your numbers, and multiply the non-zero numbers.

80,000 → 8   50 → 5

Step 2: At the end of your product, replace the total number of zeros you removed from the multipliers. Because you removed a total of five zeros from your multipliers, place five zeros after your product.

80,000 × 50 = 4,000,000

It's Your Turn!

Using the method above, find the products of the following problems, and write the corresponding letter from the correct answer on the line.

1. 300 × 90,000   A. 31,720,000
2. 45 × 8500   B. 3,524,000
3. 4400 × 7500   C. 27,000,000
4. 52,000 × 610   D. 33,000,000
5. 88,100 × 40   E. 382,500

Challenge Yourself!

A super-fast chess computer can perform 200,000,000 calculations per second. How many calculations can it perform in the 3 minutes it is allowed for each move?

DIVISION

Name _____ Date _____ Class _____

MATH SKILLS

Dividing Whole Numbers with Long Division

Long division, which is used to divide numbers of more than one digit, is really just a series of simple division, multiplication, and subtraction problems. The number that you divide is called the dividend. The number you divide the dividend by is the divisor. The answer to a division problem is called a quotient.

SAMPLE PROBLEM: Divide 564 by 12, or 12)564.

Step 1: Because you cannot divide 12 into 5, you must start by dividing 12 into 56. To do this, ask yourself, "What number multiplied by 12 comes closest to 56 without going over?" 4 × 12 = 48, so place a 4 in the quotient.

Step 2: Multiply the 4 by the divisor and place the product under the 56. Then subtract that product from 56.

Step 3: Bring the next digit down from the dividend (4), and divide this new number (84) by the divisor, as you did in Step 1. Because 12 divides into 84 seven times, write 7 in the quotient.

The quotient is **47.**

Divide It Up!

1. Fill in the blanks in the following long-division problems:

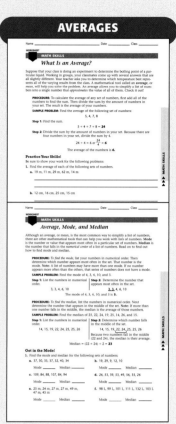

Checking Division with Multiplication

Multiplication and division "undo" one another. This means that when you ask yourself, "What is 12 divided by 3?" it is the same as asking, "What number multiplied by 3 gives 12?" You can use this method to catch mistakes in your division.

PROCEDURE: To check your division with multiplication, multiply the quotient of your division problem by the divisor and compare the result with the dividend. If they are equal, your division was correct.

SAMPLE PROBLEM: Divide 564 by 47, and check your result with multiplication.

Step 1: Divide to find your quotient.

Step 2: Multiply the quotient by the divisor.

Step 3: Compare the product with your dividend.

564 = 564 Correct!

Check It Out!

Complete the following divisions, and check your math by multiplying the quotient by your divisor. Are the product and the dividend equal?

1. 15)405
 quotient = _____   product = _____

2. 14)1694
 quotient = _____   product = _____

3. 12)252
 quotient = _____   product = _____

AVERAGES

Name _____ Date _____ Class _____

MATH SKILLS

What Is an Average?

Suppose that your class is doing an experiment to determine the boiling point of a particular liquid. Working in groups, your classmates come up with several answers that are all slightly different. Your teacher asks you to determine which temperature best represents all of the varying results from the class. A mathematical tool called an **average,** or **mean,** will help you solve the problem. An average allows you to simplify a list of numbers into a single number that approximates the value of all of them. Check it out!

PROCEDURE: To calculate the average of any set of numbers, first add all of the numbers to find the sum. Then divide the sum by the amount of numbers in your set. The result is the average of your numbers.

SAMPLE PROBLEM: Find the average of the following set of numbers: 5, 4, 7, 8

Step 1: Find the sum.

5 + 4 + 7 + 8 = 24

Step 2: Divide the sum by the amount of numbers in your set. Because there are four numbers in your set, divide the sum by 4.

24 ÷ 4 = 6 or 24/4 = 6

The average of the numbers is **6.**

Practice Your Skills!

Be sure to show your work for the following problems.

1. Find the average of each of the following sets of numbers.
 a. 19 m, 11 m, 29 m, 62 m, 14 m

 b. 12 cm, 16 cm, 25 cm, 15 cm

Average, Mode, and Median

Although an average, or mean, is the most common way to simplify a list of numbers, there are other mathematical tools that can help you work with lists of numbers. **Mode** is the number or value that appears most often in a particular set of numbers. **Median** is the number that falls in the numerical center of a list of numbers. Read on to find out how to find mode and median.

PROCEDURE: To find the mode, list your numbers in numerical order. Then determine which number appears most often in the set. That number is the mode. Note: A list of numbers may have more than one mode. If no number appears more often than the others, that series of numbers does not have a mode.

SAMPLE PROBLEM: Find the mode of 4, 3, 6, 10, and 3.

Step 1: List the numbers in numerical order.

3, 3, 4, 6, 10

Step 2: Determine the number that appears most often in the set.

3, 3, 4, 6, 10

The mode of 4, 3, 6, 10, and 3 is **3.**

PROCEDURE: To find the median, list the numbers in numerical order. Next determine the number that appears in the middle of the set. **Note:** If more than one number falls in the middle, the median is the average of those numbers.

SAMPLE PROBLEM: Find the median of 25, 22, 24, 19, 25, 14, 26, and 15.

Step 1: List the numbers in numerical order.

14, 15, 19, 22, 24, 25, 25, 26

Step 2: Determine which number falls in the middle of the set.

14, 15, 19, 22, 24, 25, 25, 26

Because two numbers fall in the middle (22 and 24), the median is their average.

Median = (22 + 24) ÷ 2 = **23**

Get in the Mode!

1. Find the mode and median for the following sets of numbers.
 a. 37, 30, 33, 37, 32, 40, 34   Mode ___ Median ___   b. 19, 29, 9, 12, 10   Mode ___ Median ___
 c. 109, 84, 88, 107, 84, 94   Mode ___ Median ___   d. 28, 53, 33, 49, 56, 15, 26   Mode ___ Median ___
 e. 25 m, 24 m, 27 m, 27 m, 49 m, 47 m, 45 m   Mode ___ Median ___   f. 98 L, 99 L, 101 L, 111 L, 132 L, 103 L   Mode ___ Median ___

POSITIVE AND NEGATIVE NUMBERS

Name _____ Date _____ Class _____

MATH SKILLS

Comparing Integers on a Number Line

An **integer** is any whole number (0, 1, 2, 3, . . .) or its opposite. You can use a number line, which is used to represent positive and negative numbers in order. A number line looks like this:

The farther a number is to the right on a number line, the greater the number. The farther a number is to the left on a number line, the smaller the number.

PROCEDURE: To compare integers on a number line, simply place your values on the line, with positive numbers to the right of zero and negative numbers to the left of zero. The number that is the farthest to the right is the greatest number. The number that is the farthest to the left is the smallest number.

SAMPLE PROBLEM: Which is greater, −8 or −3?

Step 1: Draw your number line and select a point for 0. Then fill in the integer values on the line.

Step 2: Place the integers you are comparing on the number line. Because both numbers are negative, they will both be to the left of zero.

Because −3 is farther to the right than −8, −3 is greater than −8.

Practice Your Skills!

1. Locate the following integers on the number line below. Then list them in order from smallest to greatest on the line below.

4, 12, −2, 7, −5, 2, −7, 9, −13

2. Use a number line to correctly place the sign > (greater than) or < (less than) between

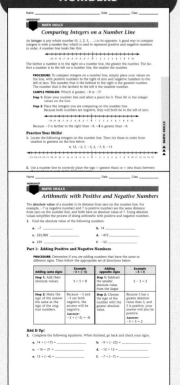

Arithmetic with Positive and Negative Numbers

The **absolute value** of a number is its distance from zero on the number line. For example, −7 (a negative number) and 7 (a positive number) are the same distance from zero on the number line, and both have an absolute value of 7. Using absolute values simplifies the process of doing arithmetic with positive and negative numbers.

1. Find the absolute value of the following numbers.
 a. −7 _____   b. 14 _____
 c. 325,000 _____   d. −475 _____
 e. 230 _____   f. −52 _____

Part 1: Adding Positive and Negative Numbers

PROCEDURE: Determine if you are adding numbers that have the same or different signs. Then follow the appropriate set of directions below.

Adding same signs	Example −3 + (−5)	Adding opposite signs	Example −3 + 5
Step 1: Add their absolute values.	3 + 5 = 8	**Step 1:** Subtract the smaller absolute value from the larger.	5 − 3 = 2
Step 2: Make the sign of the answer the same as the sign of the original numbers. Answer: −3 + (−5) = −8	Because −3 and −5 are both negative, the answer will be negative. Answer: −3 + (−5) = −8	**Step 2:** Choose the sign of the number with the greater absolute value.	Because 5 has a greater absolute value than 3, and 5 is positive, the answer will also be positive. Answer: −3 + 5 = 2

Add It Up!

2. Complete the following equations. When finished, go back and check your signs.
 a. 14 + (−17) = _____   b. −9 + (−23) = _____
 c. −16 + 21 = _____   d. −12 + 12 = _____
 e. 13 + (−4) = _____   f. −7 + (−7) = _____

FRACTIONS

Name _____ Date _____ Class _____

MATH SKILLS

What Is a Fraction?

Suppose that you are doing an experiment in your class on the benefits of sunlight to plants. Your teacher has asked you to put ¼ of the plants in the sun. What does that mean? While whole numbers, such as 1 and 879, are used to indicate how many, the number below the fraction bar in a fraction is called the **denominator.** This number indicates how many parts there are in the whole. The number above the fraction bar, called the **numerator,** tells you how many parts of that whole are represented.

PROCEDURE: To make a fraction, write the total number of units in the whole as your denominator. Then write the number of parts of that whole being represented as the numerator.

SAMPLE PROBLEM: Your class has 24 plants. Your teacher instructs you to put 5 in a shady spot. What fraction does this represent?

Step 1: Write the total number of units in the whole as the denominator.

5/24

Step 2: Write the number of parts of the whole being represented as the numerator.

5/24

5/24 of the plants will be in the shade.

Constructing Fractions

1. What fraction of the whole does the shaded or patterned part represent?
 a.   b.   c.

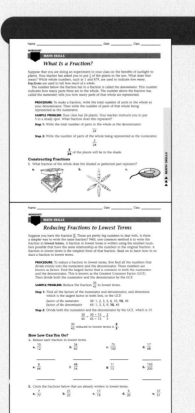

Reducing Fractions to Lowest Terms

Suppose you have the fraction 30/45. Those are pretty big numbers to deal with. Is there a simpler way to write the fraction? Well, one common method is to write the fraction in lowest terms. A fraction in lowest terms is written using the smallest numbers possible that have the same relationship as the numbers in the original fraction. A fraction in lowest terms is the simplest form of that fraction. Read on to learn how to reduce a fraction to lowest terms.

PROCEDURE: To reduce a fraction to lowest terms, first find all the numbers that divide evenly into the numerator and the denominator. These numbers are known as factors. Find the largest factor that is common to both the numerator and the denominator. This is known as the Greatest Common Factor (GCF). Then divide both the numerator and the denominator by the GCF.

SAMPLE PROBLEM: Reduce the fraction 30/45 to lowest terms.

Step 1: Find all the factors of the numerator and denominator, and determine which is the largest factor in both lists, or the GCF.

factors of the numerator 30: 1, 2, 3, 5, 6, 10, **15**, 30
factors of the denominator 45: 1, 3, 5, 9, **15**, 45

Step 2: Divide both the numerator and the denominator by the GCE, which is 15.

30/45 → 30 ÷ 15 / 45 ÷ 15 = 2/3

30/45 reduced to lowest terms is 2/3.

How Low Can You Go?

1. Reduce each fraction to lowest terms.
 a. 10/12   b. 36/60   c. 75/100   d. 17/68
 e. 8/64   f. 48/54   g. 9/15   h. 150/200

2. Circle the fractions below that are already written in lowest terms.
 a. 2/77   b. 11/25   c. 17/19   d. 9/20   e. 37/31

Improper Fractions and Mixed Numbers

An **improper fraction** is a fraction whose numerator is greater than its denominator, such as 17/5. An improper fraction can be changed to a **mixed number,** which is a whole number with a fraction, such as 2⅖. Likewise, a mixed number can be changed to an improper fraction when it is necessary for doing mathematical operations with these numbers.

PROCEDURE: To change an improper fraction to a mixed number, divide the numerator by the denominator and write the quotient as the whole number. If there is a remainder, place it over the denominator to make the fraction of the mixed number.

SAMPLE PROBLEM: Change 17/5 to a mixed number.

Step 1: Divide the numerator by the denominator.

17 ÷ 5 = 3, remainder 2

Step 2: Write the quotient as the whole number, and put the remainder over the original denominator as the fraction.

17/5 = 3 2/5

PROCEDURE: To change a mixed number to an improper fraction, multiply the denominator of the fraction by the whole number. Then add that product to the numerator. Finally, write the sum over the denominator.

SAMPLE PROBLEM: Change 4⅔ to an improper fraction.

Step 1: Multiply the denominator by the whole number.

3 × 4 = 12

Step 2: Add the product to the numerator, and write the sum over the denominator.

12 + 2 = 14   4⅔ = 14/3

1. Write True or False next to each equation.
 a. 3 1/3 = 9/3   b. 23/4 = 5 3/4

Adding and Subtracting Fractions

Part 1: Adding and Subtracting Fractions with the Same Denominator

PROCEDURE: To add fractions with the same denominator, add the numerators and put the sum over the original denominator. To subtract fractions with the same denominator, subtract the numerators and put the difference over the original denominator.

SAMPLE PROBLEM A: 3/5 + 1/5 = ?

Add the numerators, and put the sum over the original denominator:

3/5 + 1/5 = 3 + 1 / 5 = 4/5

SAMPLE PROBLEM B: 8/11 − 3/11 = ?

Subtract the numerators and put the difference over the original denominator:

8/11 − 3/11 = 8 − 3 / 11 = 5/11

Practice What You've Learned!

1. Add and subtract to complete the following equations. Reduce your answers to lowest terms.
 a. 9/17 + 6/17 =   b. 8/24 + 4/24 =
 c. 3/4 − 1/4 =   d. 16/22 − 7/22 =

Part 2: Adding and Subtracting Fractions with Different Denominators

Sometimes you have to add or subtract fractions that have different denominators. To do this, you first need to rewrite your fractions so that they DO have the same denominator. Figuring out the **least common denominator (LCD)** of your fractions is the first step.

PROCEDURE: To find the least common denominator of two fractions, find the least common multiple of the denominators. In other words, look at the multiples of the numbers, and find out which they have in common. The common multiple with the lowest value is your LCD.

SAMPLE PROBLEM: What is the LCD of 3/4 and 2/3?

Step 1: List the multiples of 4.
(4 × 1) = 4, (4 × 2) = 8, (4 × 3) = **12**, (4 × 4) = 16, etc.

Step 2: List the multiples of 3
(3 × 1) = 3, (3 × 2) = 6, (3 × 3) = 9, (3 × 4) = **12**, etc.

The least common denominator of 3/4 and 2/3 is **12.**

Multiplying and Dividing Fractions

Compared with adding and subtracting fractions, multiplying and dividing fractions is quite simple. Just follow the steps below to see how it is done.

PROCEDURE: To multiply fractions, multiply the numerators and the denominators together and reduce the fraction (if necessary).

SAMPLE PROBLEM A: 5/9 × 7/10 = ?

Step 1: Multiply the numerators and denominators.

5/9 × 7/10 = 5 × 7 / 9 × 10 = 35/90

Step 2: Reduce.

35/90 = 35 ÷ 5 / 90 ÷ 5 = 7/18

Answer:

5/9 × 7/10 = 7/18

PROCEDURE: To divide fractions, switch the numerator and denominator of the divisor (the number you divide by) to make that fraction's reciprocal. Then multiply the fraction and the reciprocal, and reduce if necessary.

SAMPLE PROBLEM B: 5/8 ÷ 2/3 = ?

Step 1: Rewrite the divisor as its reciprocal.

2/3 → 3/2

Step 2: Multiply the dividend by the reciprocal.

5/8 × 3/2 = 5 × 3 / 8 × 3 = 15/24

Step 3: Reduce.

15/24 = 15 ÷ 3 / 24 ÷ 3 = 5/12

Practice Your Skills!

1. Multiply and divide to complete the equations. Give your answers in lowest terms.
 a. 4/5 × 2/3 =   b. 1/2 ÷ 7/8 =
 c. 6/7 × 3/12 =   d. 1 1/4 ÷ 3/8 =

2. You have 23¼ L of saline solution. Every student in the class needs 1⅛ L for an experiment. How many students can do the experiment?

3. Because of differences in gravity, your weight on the moon would be ⅙ what it is on Earth. If you weigh 72 N, what would be your weight on the moon?

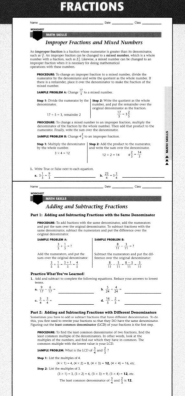
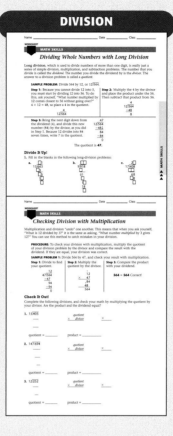

Math Skills for Science (continued)

RATIOS AND PROPORTIONS

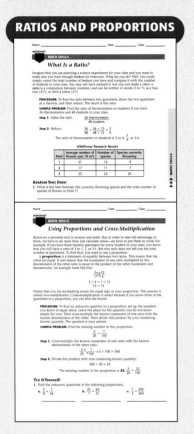

DECIMALS

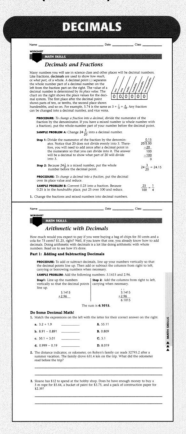

PERCENTAGES

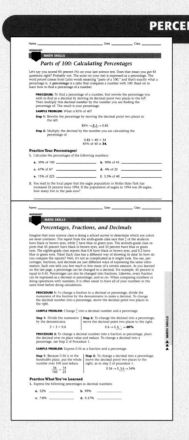

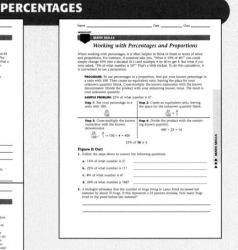

POWERS OF 10

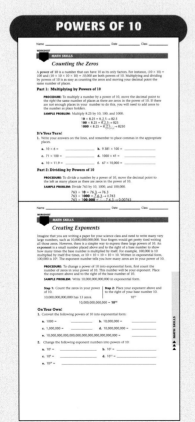

SCIENTIFIC NOTATION

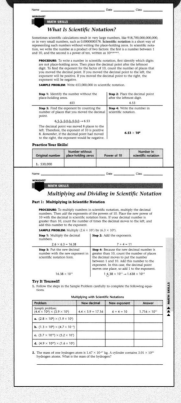

SI MEASUREMENT AND CONVERSION

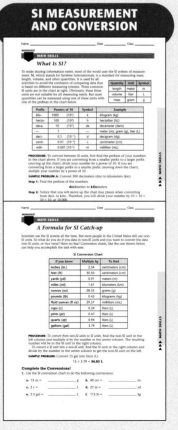

Math Skills for Science (continued)

GEOMETRY

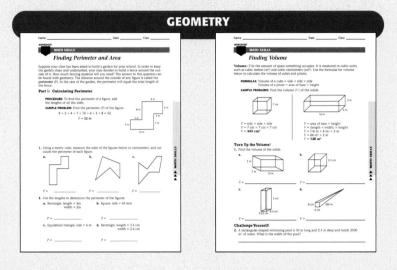

THE UNIT FACTOR AND DIMENSIONAL ANALYSIS

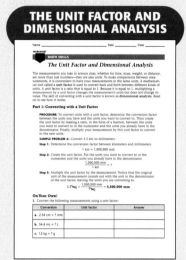

MATH IN SCIENCE: INTEGRATED SCIENCE

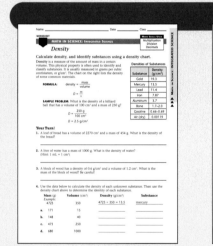

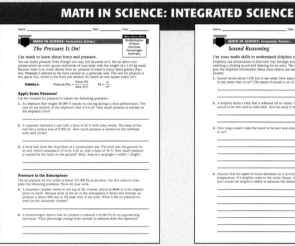

Science & Math Skills Worksheets (continued)

Math Skills for Science (continued)

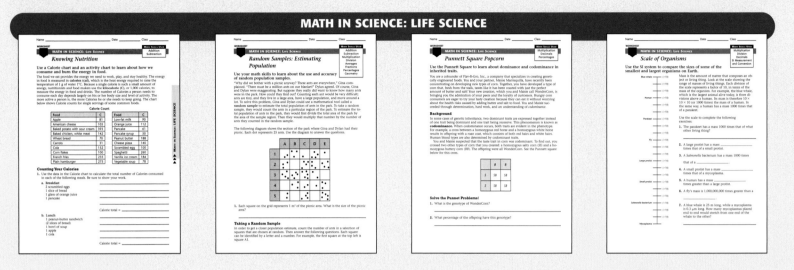

Assessment Checklist & Rubrics

The following is just a sample of over 50 checklists and rubrics contained in this booklet.

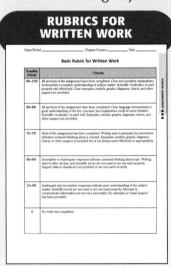

RUBRICS FOR WRITTEN WORK

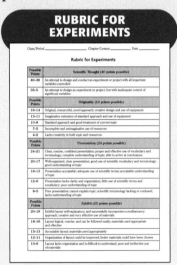

RUBRIC FOR EXPERIMENTS

TEACHER EVALUATION OF COOPERATIVE LEARNING

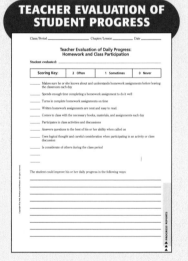

TEACHER EVALUATION OF STUDENT PROGRESS

LIFE SCIENCE

NATIONAL SCIENCE EDUCATION STANDARDS CORRELATIONS

The following lists show the chapter correlation of **Holt Science and Technology: Environmental Science** with the *National Science Education Standards* (grades 5-8)

UNIFYING CONCEPTS AND PROCESSES

Standard	Chapter Correlation	
Systems, order, and organization Code: UCP 1	Chapter 1 Chapter 2	1.1, 1.2, 1.3 2.1, 2.2
Evidence, models, and explanation Code: UCP 2	Chapter 2 Chapter 3 Chapter 4 Chapter 5	2.1, 2.2 3.1, 3.2, 3.3 4.2 5.3
Change, constancy, and measurement Code: UCP 3	Chapter 2 Chapter 3 Chapter 4 Chapter 5	2.1, 2.2 3.2, 3.3 4.1, 4.2 5.3
Evolution and equilibrium Code: UCP 4	Chapter 1 Chapter 2 Chapter 3 Chapter 4	1.3 2.1, 2.2 3.1, 3.3 4.2
Form and function Code: UCP 5	Chapter 3 Chapter 5	3.1 5.3

SCIENCE AND TECHNOLOGY

Standard	Chapter Correlation	
Abilities of technological design Code: ST 1	Chapter 3 Chapter 5	3.1 5.3
Understandings about science and technology Code: ST 2	Chapter 4 Chapter 5	4.2 5.3

HISTORY AND NATURE OF SCIENCE

Standard	Chapter Correlation	
Nature of science Code: HNS 2	Chapter 4	4.2
History of science Code: HNS 3	Chapter 4	4.1

SCIENCE AS INQUIRY

Standard	Chapter Correlation	
Abilities necessary to do scientific inquiry Code: SAI 1	Chapter 1 Chapter 2 Chapter 3 Chapter 4 Chapter 5	1.1, 1.2 2.1, 2.2 3.1, 3.2, 3.3 4.1, 4.2 5.1, 5.2, 5.3
Understandings about scientific inquiry Code: SAI 2	Chapter 1 Chapter 2 Chapter 3 Chapter 4	1.1 2.1 3.2 4.1

SCIENCE IN PERSONAL AND SOCIAL PERSPECTIVES

Standard	Chapter Correlation	
Personal health Code: SPSP 1	Chapter 4 Chapter 5	4.1, 4.2 5.2
Populations, resources, and environments Code: SPSP 2	Chapter 1 Chapter 2 Chapter 3 Chapter 4 Chapter 5	1.1 2.2 3.1, 3.2, 3.3 4.1, 4.2 5.1, 5.2, 5.3
Natural hazards Code: SPSP 3	Chapter 3 Chapter 4	3.1 4.1
Risks and benefits Code: SPSP 4	Chapter 2 Chapter 4 Chapter 5	2.1 4.1, 4.2 5.1, 5.2, 5.3
Science and technology in society Code: SPSP 5	Chapter 2 Chapter 4 Chapter 5	2.1 4.1, 4.2 5.2, 5.3

LIFE SCIENCE NATIONAL SCIENCE EDUCATION CONTENT STANDARDS

STRUCTURE AND FUNCTION IN LIVING SYSTEMS

Standard	Chapter Correlation	
Living systems at all levels of organization demonstrate the complementary nature of structure and function. Important levels of organization for structure and function include cells, organs, tissues, organ systems, whole organisms, and ecosystems. Code: LS 1a	**Chapter 1** **Chapter 3**	1.1, 1.2, 1.3 3.1, 3.2, 3.3

REGULATION AND BEHAVIOR

Standard	Chapter Correlation	
All organisms must be able to obtain and use resources, grow, reproduce, and maintain stable internal conditions while living in a constantly changing external environment. Code: LS 3a	**Chapter 1** **Chapter 3** **Chapter 4**	1.3 3.1 4.1
Behavior is one kind of response an organism can make to an internal or environmental stimulus. A behavioral response requires coordination and communication at many levels, including cells, organ systems, and whole organisms. Behavioral response is a set of actions determined in part by heredity and in part from experience. Code: LS 3c	**Chapter 1** **Chapter 3**	1.3 3.1
An organism's behavior evolves through adaptation to its environment. How a species moves, obtains food, reproduces, and responds to danger are based in the species' evolutionary history. Code: LS 3d	**Chapter 1** **Chapter 3**	1.2 3.1, 3.2, 3.3

POPULATIONS AND ECOSYSTEMS

Standard	Chapter Correlation	
A population consists of all individuals of a species that occur together at a given place and time. All populations living together and the physical factors with which they interact compose an ecosystem. Code: LS 4a	**Chapter 1** **Chapter 3**	1.1, 1.2, 1.3 3.2, 3.3
Populations of organisms can be categorized by the functions they serve in an ecosystem. Plants and some microorganisms are producers—they make their own food. All animals, including humans, are consumers, which obtain food by eating other organisms. Decomposers, primarily bacteria and fungi, are consumers that use waste materials and dead organisms for food. Food webs identify the relationships among producers, consumers, and decomposers in an ecosystem. Code: LS 4b	**Chapter 1** **Chapter 3**	1.2 3.1, 3.2, 3.3
For ecosystems, the major source of energy is sunlight. Energy entering ecosystems as sunlight is transferred by producers into chemical energy through photosynthesis. That energy then passes from organism to organism in food webs. Code: LS 4c	**Chapter 1** **Chapter 3**	1.1, 1.2 3.1, 3.2, 3.3
The number of organisms an ecosystem can support depends on the resources available and abiotic factors, such as quantity of light and water, range of temperatures, and soil composition. Given adequate biotic and abiotic resources and no disease or predators, populations (including humans) increase at rapid rates. Lack of resources and other factors, such as predation and climate, limit the growth of populations in specific niches in the ecosystem. Code: LS 4d	**Chapter 1** **Chapter 3** **Chapter 4**	1.1, 1.2, 1.3 3.1, 3.2, 3.3 4.1

DIVERSITY AND ADAPTATIONS OF ORGANISMS

Standard	Chapter Correlation
Millions of species of animals, plants, and microorganisms are alive today. Although different species might look dissimilar, the unity among organisms becomes apparent from an analysis of internal structures, the similarity of their chemical processes, and the evidence of common ancestry. Code: LS 5a	**Chapter 3** 3.1
Biological evolution accounts for the diversity of species developed through gradual processes over many generations. Species acquire many of their unique characteristics through biological adaptation, which involves the selection of naturally occurring variations in populations. Biological adaptations include changes in structures, behaviors, or physiology that enhance survival and reproductive success in a particular environment. Code: LS 5b	**Chapter 3** 3.1
Extinction of a species occurs when the environment changes and the adaptive characteristics of a species are insufficient to allow its survival. Fossils indicate that many organisms that lived long ago are extinct. Extinction of species is common; most of the species that have lived on Earth no longer exist. Code: LS 5c	**Chapter 4** 4.2

STRUCTURE OF THE EARTH SYSTEM

Standard	Chapter Correlation
Living organisms have played many roles in the earth system, including affecting the composition of the atmosphere, producing some types of rocks, and contributing to the weathering of rocks. Code: ES 1k	**Chapter 5** 5.2

EARTH IN THE SOLAR SYSTEM

Standard	Chapter Correlation
The sun is the major source of energy for phenomena on the earth's surface, such as growth of plants, winds, ocean currents, and the water cycle. Seasons result from variations in the amount of the sun's energy hitting the surface, due to the tilt of the earth's rotation on its axis and the length of the day. Code: ES 3d	**Chapter 5** 5.2, 5.3

Master Materials List

For added convenience, Science Kit® provides materials-ordering software on CD-ROM designed specifically for *Holt Science and Technology*. Using this software, you can order complete kits or individual items, quickly and efficiently.

CONSUMABLE MATERIALS	AMOUNT	PAGE
Aluminum foil	1 sheet	136
Animal, aquatic	5	29
Bag, paper lunch	1	128
Balloon, round, black	1	97
Balloon, round, colored	3	97
Balloon, round, white	1	97
Beans, pinto	4–12 oz.	128
Bottle, soda, 2 L	1	47, 116
Candle	1	32
Card, index, 3 x 5 in.	1	116
Card, index, 3 x 5 in.	4	3
Card, index, 3 x 5 in.	20	61
Clay, modeling	1 stick	32, 116, 136
Fertilizer	1/4 cup	64
Gloves, protective	1 pair	38, 64, 130, 136
Glue, white	1 bottle	116
Gravel	6 oz	47
Ice cubes, large	5–10	97
Jug, plastic, milk	1	116
Limestone	1 sample	103
Magazine	2	20, 88
Marker, black, permanent	1	116, 128, 136
Marker, various colors	1 pack	20

CONSUMABLE MATERIALS	AMOUNT	PAGE
Newspaper	1	73, 88
Oil, machine (light)	15 drops	103
Paper, filter	2 pieces	130
Paper, graph	1 sheet	5
Paper, pH	1 piece	130
Pencil, wax	1	64
Plant, small various species	3	47
Plant, various species	5	29
Plastic wrap, clear, approx. 1 x 2 ft	1 sheet	38, 64
Pond water with living organisms	300 mL	64
Poster board	1	20
Safety razor (for teacher)	1	116
Sandstone	1 sample	103
Shale	1 sample	103
Skewer, wooden	2	116
Soil	8 oz	47
Soil	500 g	38
Soil, potting (without fertilizer)	4 cups	130
String (or yarn)	3 m	61
Tape, transparent	1 roll	116
Thread, 20 cm	1	116
Water, distilled	60 mL	130
Water, distilled	2.25 L	64

Nonconsumable Equipment	Amount	Page
Aquarium	1	29
Balance	1	38, 130, 132
Beaker, 50 mL	1	130
Beaker, 1 L	1	73
Blender	1	73
Board, flat, 0.5 x 0.5 m	1	73
Calculator	1	128, 136
Coin	1	116
Cork, small	1	116
Coverslip, plastic	1	64
Eyedropper, plastic	1	64
Fishbowl, large	1	38
Funnel, glass	1	130
Graduated cylinder, 25 mL	1	130
Graduated cylinder, 100 mL	1	64
Graduated cylinder, 250 mL	1	38
Hole punch	1	116
Insect, large, dead	5	130

Nonconsumable Equipment	Amount	Page
Jar, 1 pt, with lid	1	130
Jar, 1 qt	3	64
Jar, glass, with lid	1	32, 136
Jar, wide-mouthed	1	47
Lamp, goose-neck, 100 W bulb	1	136
Microscope, compound	1	64
Microscope slide, plastic	1	64
Pan, square	1	73
Petri dish, plastic	3	103
Ruler, metric	1	5, 116, 136
Scissors	1	20, 116
Sponge	2	132
Stirring rod	1	64
Stopwatch	1	116, 136
Thermometer, Celsius	1	136
Thumbtack	5	116
Wire screen, approx. 2 x 2 ft	1	73

Answers to Concept Mapping Questions

The following pages contain sample answers to all of the concept mapping questions that appear in the Chapter Reviews. Because there is more than one way to do a concept map, your students' answers may vary.

CHAPTER 1 Interactions of Living Things

18.

The biosphere

↓ includes

ecosystems

↓ composed of

communities

↓ composed of

populations

↓ composed of

individual organisms

↓ composed of

producers   consumers

which include

herbivores   carnivores

CHAPTER 2 Cycles in Nature

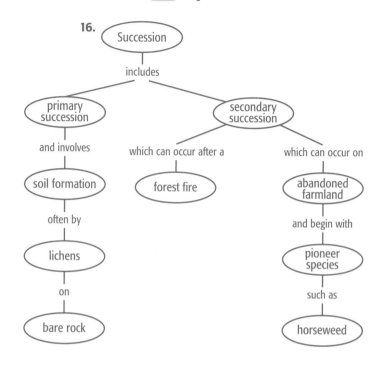

16.

Succession

↓ includes

primary succession   secondary succession

and involves   which can occur after a   which can occur on

soil formation   forest fire   abandoned farmland

often by   and begin with

lichens   pioneer species

on   such as

bare rock   horseweed

CHAPTER 3 The Earth's Ecosystems

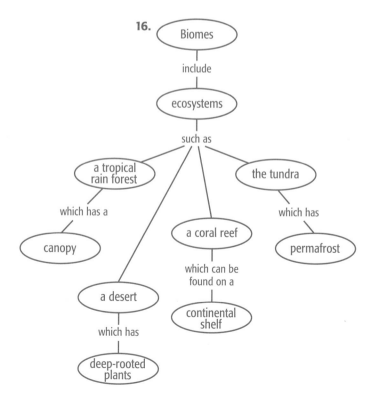

16.

Biomes

↓ include

ecosystems

↓ such as

a tropical rain forest   the tundra

which has a   a coral reef   which has

canopy   permafrost

a desert   which can be found on a

which has   continental shelf

deep-rooted plants

14.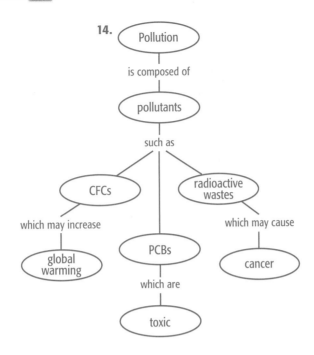

CHAPTER **5** **Energy Resources**

18.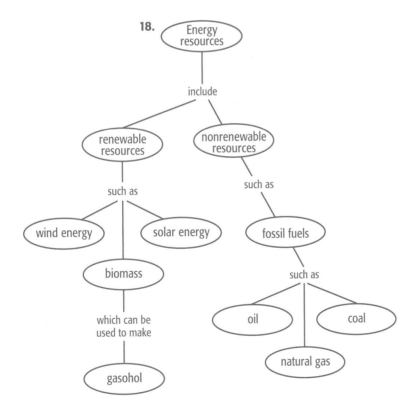

To the Student

This book was created to make your science experience interesting, exciting, and fun!

Go for It!

Science is a process of discovery, a trek into the unknown. The skills you develop using *Holt Science & Technology*— such as observing, experimenting, and explaining observations and ideas— are the skills you will need for the future. There is a universe of exploration and discovery awaiting those who accept the challenges of science.

Science & Technology

You see the interaction between science and technology every day. Science makes technology possible. On the other hand, some of the products of technology, such as computers, are used to make further scientific discoveries. In fact, much of the scientific work that is done today has become so technically complicated and expensive that no one person can do it entirely alone. But make no mistake, the creative ideas for even the most highly technical and expensive scientific work still come from individuals.

Activities and Labs

The activities and labs in this book will allow you to make some basic but important scientific discoveries on your own. You can even do some exploring on your own at home! Here's your chance to use your imagination and curiosity as you investigate your world.

Keep a ScienceLog

In this book, you will be asked to keep a type of journal called a ScienceLog to record your thoughts, observations, experiments, and conclusions. As you develop your ScienceLog, you will see your own ideas taking shape over time. You'll have a written record of how your ideas have changed as you learn about and explore interesting topics in science.

Know "What You'll Do"

The "What You'll Do" list at the beginning of each section is your built-in guide to what you need to learn in each chapter. When you can answer the questions in the Section Review and Chapter Review, you know you are ready for a test.

Check Out the Internet

You will see this *sciLINKS* logo throughout the book. You'll be using *sciLINKS* as your gateway to the Internet. Once you log on to *sciLINKS* using your computer's Internet link, type in the *sciLINKS* address. When asked for the keyword code, type in the keyword for that topic. A wealth of resources is now at your disposal to help you learn more about that topic.

In addition to *sciLINKS* you can log on to some other great resources to go with your text. The addresses shown below will take you to the home page of each site.

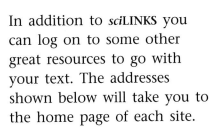

internet connect

This textbook contains the following on-line resources to help you make the most of your science experience.

go.hrw.com

Visit **go.hrw.com** for extra help and study aids matched to your textbook. Just type in the keyword HG2 HOME.

SCiLINKS NSTA

Visit **www.scilinks.org** to find resources specific to topics in your textbook. Keywords appear throughout your book to take you further.

Smithsonian Institution®
Internet Connections

Visit **www.si.edu/hrw** for specifically chosen on-line materials from one of our nation's premier science museums.

CNN fyi.com

Visit **www.cnnfyi.com** for late-breaking news and current events stories selected just for you.

Chapter Organizer

CHAPTER ORGANIZATION	TIME MINUTES	OBJECTIVES	LABS, INVESTIGATIONS, AND DEMONSTRATIONS
Chapter Opener pp. 2–3	45	National Standards: UCP 1, 5, SAI 1	**Start-Up Activity,** Who Eats Whom? p. 3
Section 1 **Everything Is Connected**	90	▶ Distinguish between the biotic and abiotic environment. ▶ Explain how populations, communities, ecosystems, and the biosphere are related. ▶ Explain how the abiotic environment relates to communities. UCP 1, SAI 1, SPSP 2, LS 1a, 4a, 4c, 4d; Labs SAI 2, LS 4d	**QuickLab,** The Human Population, p. 5 **Skill Builder,** Capturing the Wild Bean, p. 128 **Datasheets for LabBook,** Capturing the Wild Bean
Section 2 **Living Things Need Energy**	90	▶ Describe the functions of producers, consumers, and decomposers in an ecosystem. ▶ Distinguish between a food chain and a food web. ▶ Explain how energy flows through a food web. ▶ Distinguish between an organism's habitat and its niche. UCP 1, SAI 1, LS 1a, 4a–4d	**EcoLabs & Field Activities,** Survival Is Just a Roll of the Dice **Whiz-Bang Demonstrations,** Voracious Fly Catcher
Section 3 **Types of Interactions**	135	▶ Distinguish between the two types of competition. ▶ Give examples of predators and prey. ▶ Distinguish between mutualism, commensalism, and parasitism. ▶ Define *coevolution,* and give an example. UCP 1, 4, LS 3a, 3c, 4a, 4d; Labs UCP 1, 4, SAI 1, LS 1a, 3d	**Interactive Explorations CD-ROM,** What's Bugging You? A *Worksheet* is also available in the *Interactive Explorations Teacher's Edition.* **Making Models,** Adaptation: It's a Way of Life, p. 20 **Datasheets for LabBook,** Adaptation: It's a Way of Life **Long-Term Projects & Research Ideas,** Out of House and Home

---

*See page **T23** for a complete correlation of this book with the*

NATIONAL SCIENCE EDUCATION STANDARDS.

TECHNOLOGY RESOURCES

 Guided Reading Audio CD
English or Spanish, Chapter 1

 One-Stop Planner CD-ROM with Test Generator

 Interactive Explorations CD-ROM
CD 2, Exploration 1, What's Bugging You?

 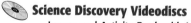 **Eye on the Environment,** Hummingbird Mites, Segment 3

 Science Discovery Videodiscs
Image and Activity Bank with Lesson Plans: Balanced Ecosystems

Science Sleuths: Neo-Cassava: The Tropical Miracle

CLASSROOM WORKSHEETS, TRANSPARENCIES, AND RESOURCES	SCIENCE INTEGRATION AND CONNECTIONS	REVIEW AND ASSESSMENT
Directed Reading Worksheet **Science Puzzlers, Twisters & Teasers**		
Transparency 65, The Five Levels of Environmental Organization **Directed Reading Worksheet,** Section 1	**Multicultural Connection,** p. 5 in ATE **Multicultural Connection,** p. 6 in ATE **Connect to Environmental Science,** p. 6 in ATE **Eye on the Environment:** Alien Invasion, p. 27	**Section Review,** p. 7 **Quiz,** p. 7 in ATE **Alternative Assessment,** p. 7 in ATE
Transparency 116, Formation of Coal **Directed Reading Worksheet,** Section 2 **Math Skills for Science Worksheet,** Working with Percentages and Proportions **Transparency 66,** Energy Pyramid **Reinforcement Worksheet,** Weaving a Food Web **Critical Thinking Worksheet,** A Struggle to Survive	**Connect to Earth Science,** p. 8 in ATE **Real-World Connection,** p. 9 in ATE **Math and More,** p. 10 in ATE **MathBreak,** Energy Pyramids, p. 11	**Self-Check,** p. 9 **Self-Check,** p. 10 **Section Review,** p. 13 **Quiz,** p. 13 in ATE **Alternative Assessment,** p. 13 in ATE
Directed Reading Worksheet, Section 3 **Reinforcement Worksheet,** Symbiotic Relationships	**Real-World Connection,** p. 17 in ATE **Multicultural Connection,** p. 18 in ATE **Apply,** p. 19 **Health Watch:** An Unusual Guest, p. 26	**Self-Check,** p. 15 **Section Review,** p. 19 **Quiz,** p. 19 in ATE **Alternative Assessment,** p. 19 in ATE

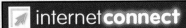 internetconnect

 Holt, Rinehart and Winston On-line Resources

go.hrw.com

For worksheets and other teaching aids related to this chapter, visit the HRW Web site and type in the keyword: **HSTINT**

 National Science Teacher's Association

www.scilinks.org

Encourage students to use the *sci*LINKS numbers listed in the internet connect boxes to access information and resources on the **NSTA** Web site.

END-OF-CHAPTER REVIEW AND ASSESSMENT

Chapter Review in Study Guide
Vocabulary and Notes in Study Guide
Chapter Tests with Performance-Based Assessment, Chapter 1 Test
Chapter Tests with Performance-Based Assessment, Performance-Based Assessment 1
Concept Mapping Transparency 18

Chapter Resources & Worksheets

Visual Resources

TEACHING TRANSPARENCIES

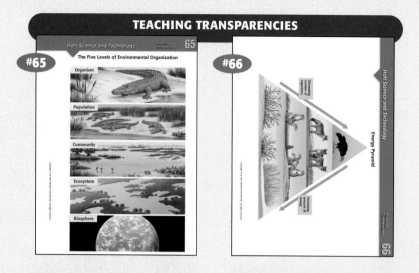

#65 — The Five Levels of Environmental Organization: Organism, Population, Community, Ecosystem, Biosphere

#66 — Energy Pyramid

TEACHING TRANSPARENCIES

#116 — Formation of Coal

LINK TO EARTH SCIENCE

CONCEPT MAPPING TRANSPARENCY

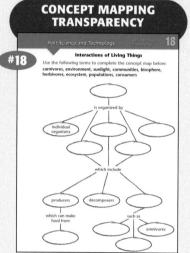

#18 — Interactions of Living Things

Use the following terms to complete the concept map below:
carnivores, environment, sunlight, communities, biosphere, herbivores, ecosystem, populations, consumers

Meeting Individual Needs

DIRECTED READING

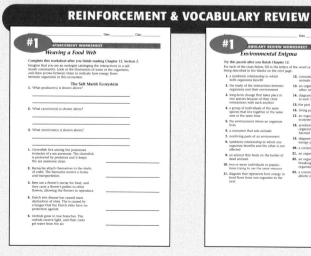

#1 — DIRECTED READING WORKSHEET

Interactions of Living Things

Chapter Introduction

As you begin this chapter, answer the following.
1. Read the title of the chapter. List three things that you already know about this subject.

2. Write two questions about this subject that you would like answered by the time you finish this chapter.

3. How does the title of the Start-Up Activity relate to the subject of the chapter?

Section 1: Everything Is Connected (p. 4)
4. In what ways do alligators provide shelter for other organisms? (Circle all that apply.)
   a. Fish live in abandoned holes dug by alligators.
   b. Tiny fish swim in and out of an alligator's mouth.
   c. Birds nest in trees on islands developed from alligator nest mounds.
   d. Alligators protect gars from predators by allowing the gars to swim beside them.

Studying the Web of Life (p. 4)
5. The study of the interactions between _____ and their _____ is called ecology.

REINFORCEMENT & VOCABULARY REVIEW

#1 — REINFORCEMENT WORKSHEET

Weaving a Food Web

Complete this worksheet after you finish reading Chapter 12, Section 2.
Imagine that you are an ecologist cataloging the interactions in a salt marsh community. Look at the illustration of some of the organisms, and draw arrows between them to indicate how energy flows between organisms in this ecosystem.

The Salt Marsh Ecosystem

1. What producer(s) is shown above?

2. What carnivore(s) is shown above?

3. What omnivore(s) is shown above?

1. Clownfish live among the poisonous tentacles of a sea anemone. The clownfish is protected by predators and it keeps the sea anemone clean.

2. Barnacles attach themselves to the shells of crabs. The barnacles receive a home and transportation.

3. Bees use a flower's nectar for food, and they carry a flower's pollen to other flowers, allowing the flowers to reproduce.

4. Dutch elm disease has caused mass destruction of elms. It is caused by a fungus that the Dutch elms have no protection against.

5. Orchids grow in tree branches. The orchids receive light, and their roots get water from the air.

#1 — VOCABULARY REVIEW WORKSHEET

Environmental Enigma

Try this puzzle after you finish Chapter 12.
For each of the clues below, fill in the letters of the word or phrase being described in the blanks on the next page.

1. a symbiotic relationship in which both organisms benefit
2. the study of the interactions between organisms and their environment
3. long-term change that takes place in two species because of their close interactions with each other
4. a group of individuals of the same species that live together in the same area at the same time
5. the environment where an organism lives
6. a consumer that eats animals
7. nonliving parts of an environment
8. symbiotic relationship in which one organism benefits and the other is not affected
9. an animal that feeds on the bodies of dead animals
10. two or more individuals or populations trying to use the same resource
11. diagram that represents how energy in food flows from one organism to the next

12. consumer that eats both plants and animals
13. an organism that eats producers or other organisms for energy
14. diagram that shows the loss of energy at each level of a food chain
15. the part of the Earth where life exists
16. living parts of an environment
17. an organism's way of life within an ecosystem
18. symbiotic relationship in which one organism benefits while the other is harmed
19. diagram that represents the many energy pathways in a real ecosystem
20. a consumer that eats plants
21. an organism that eats prey
22. an organism that gets energy by breaking down the remains of dead organisms
23. a community of organisms and their abiotic environment

SCIENCE PUZZLERS, TWISTERS & TEASERS

#1 — SCIENCE PUZZLERS, TWISTERS & TEASERS

Interactions of Living Things

Nature's Philosophers

1. A group of astronauts has discovered amazing new life-forms on a distant moon. Each life-form was asked to describe its philosophy of life in a sentence or two. Help the astronauts categorize each as a producer, carnivore, omnivore, scavenger, or decomposer by writing the appropriate term in the space provided.
   a. "To kill is wrong. To let food go to waste is worse."

   b. "The swift receive the most satisfying reward."

   c. "One must lead a balanced life."

   d. "Self-reliance is the path to contentment. Always seek the light of truth."

   e. "Ashes to ashes; dust to dust."

Find the Oddball

2. In each group of terms below, three terms are related. Circle the term that doesn't belong and explain why.
   a. grass, wind, rain, soil

   b. salamander, water lily, humidity, bacteria

   c. population, community, organism, predator

Chapter 1 • Interactions of Living Things

Review & Assessment

STUDY GUIDE

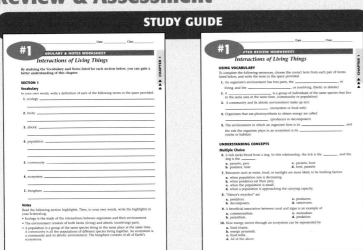

CHAPTER TESTS WITH PERFORMANCE-BASED ASSESSMENT

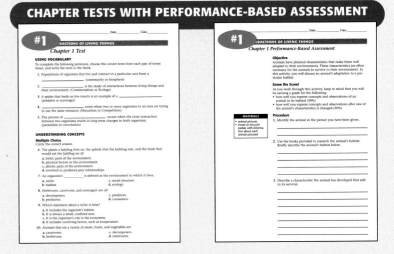

Lab Worksheets

ECOLABS & FIELD ACTIVITIES

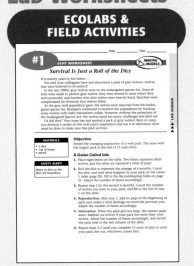

WHIZ-BANG DEMONSTRATIONS

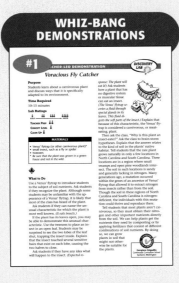

LONG-TERM PROJECTS & RESEARCH IDEAS

DATASHEETS FOR LABBOOK

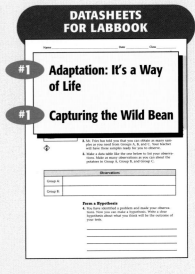

Applications & Extensions

CRITICAL THINKING & PROBLEM SOLVING

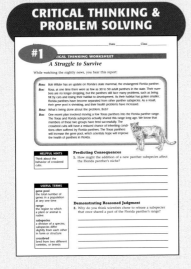

EYE ON THE ENVIRONMENT

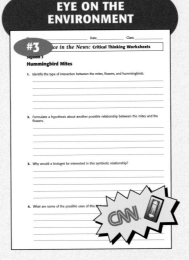

INTERACTIVE EXPLORATIONS

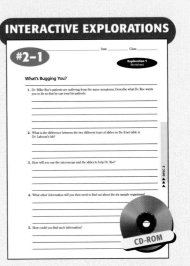

SECTION 1

Everything Is Connected

▶ Indicator Species

An indicator species is a plant or animal that, by its presence in a particular area, indicates the environmental conditions of the site. This information tells scientists what other species might thrive in the locale. For example, mosses usually indicate acidic soil, which would limit the types of plants that can survive in the same area. This, in turn, affects the variety of herbivores and other animals that can live there.

▶ Ecology

Ecology as an academic discipline can be traced to Theophrastus (c. 372–287 B.C.), a student of Aristotle (384–322 B.C.). Theophrastus offered the first descriptions of the relationships between organisms and the living and nonliving parts of their environment. The term *ecology* was coined in 1866 by Ernst Haeckel (1834–1919), a German zoologist. *Oekologie*, the German word for ecology, comes from the Greek *oikos*, which means "household, home, or place to live."

IS THAT A FACT!

- 🖝 The Australian mallee fowl regulates the temperature of its nest mound with the help of other organisms. A healthy population of fungus and bacteria inhabit the leaf-and-twig mound built by the birds. As the bacteria decompose the leaves, they give off heat. When the temperature reaches 34°C (the birds can tell by sticking their beak into the mound), it's egg-laying time! The birds work hard to keep the mound and their eggs at a constant temperature by adding more material or digging ventilation shafts. They can also fight off potential predators while their eggs remain warm in the mound.

SECTION 2

Living Things Need Energy

▶ Systems Ecology

Some early ecologists focused on communities and populations, but others looked at the energy transfer between organisms. Today the study of energy transfer is called systems ecology. It involves analysis of the flow of energy and the recycling of nutrients within an ecosystem to answer the question, "How does the ecosystem function?" The use of modern materials and techniques, such as radioisotopes, computer science, and applied mathematics, has enabled scientists to quantify the movement of nutrients and energy through ecosystems.

IS THAT A FACT!

- 🖝 The coral reef is the most productive habitat on Earth.

- 🖝 Kelp is a marine alga, often called seaweed, anchored to the ocean floor. It is a primary producer. Sea urchins and many other organisms eat it. Kelp also provides shelter for animals such as bronze kelp perch. Sea otters sometimes wrap themselves in a blade of kelp to keep from drifting while they nap.

- 🖝 Blades of the Pacific giant kelp can grow up to 60 m, taller than the Statue of Liberty.

- 🖝 Some animals build a home within their habitat. The trapdoor spider digs a burrow in the ground and constructs a "trap door" of silk and mud with silk hinges. It waits until an insect walks by, then quickly opens the door and grabs its prey. Trapdoor spiders inject venom into their prey, but they are harmless to people.

- Different habitats provide different materials for animals to use when constructing their home. These homes have different names. A squirrel's nest of leaves and sticks is called a drey. A badger's burrow is called a sett. A river otter's burrow in a river bank is called a holt.

SECTION 3

Types of Interactions

▶ Mimicry

Batesian mimicry was named for Henry Walter Bates (1825–1892), who described it in 1862. The mimic assumes the form of its model to take advantage of the model's defenses. For example, the viceroy butterfly looks just like the unappetizing monarch, which birds avoid. The snake caterpillar's movements resemble those of a real snake and thus help prevent attack by predators fearful of snakes.

- In 1878, Fritz Müller (1821–1897) described Müllerian mimicry, in which the resemblance of two species gives them mutual defense benefits. Both the sand wasp and the yellow jacket can sting. A predator that avoids one will avoid the other.

- Camouflage, a prey adaptation illustrated by the praying mantis's resemblance to a leaf, is a form of mimicry.

▶ Joke's on You: Prey Adaptations

The killdeer is a ground-nesting bird that will distract a predator from its nest or chicks with a "broken-wing" display. It will limp and drag a wing on the ground, making itself appear to be an easy catch. But because it is actually quite healthy, it always stays one step ahead of the predator.

▶ Plants and Ants

Symbiotic relationships between animals and plants are often a marvel to behold. There are epiphytes (plants that grow without soil) that have coevolved with ants. The plants provide knobby, chambered tubers in which the ants live. The excrement from the ants provides vital nutrition for the plant.

- Some tree-dwelling tropical ants collect and plant seeds in gardens they tend in their tree homes. The ants nourish the seeds with feces they collect and bring to the garden. The plants then grow and provide the ants with food.

- Other ants live inside the stems of rattan palms. When one of these palms, the ant-filled *Korthalsia*, is touched, the ants inside the plant begin smashing their mandibles (jaws) together, creating a hissing, whispering, or rattling noise. No one is quite sure of the function of this activity, but it may serve to deter interested animals from further disturbing the plant.

- In West Africa, stinging ants protect the *Barteria*, the small tree in which the ants live. The sting of this ant can numb a human for several days and can penetrate even the tough skin of an elephant.

▶ Mutualism

Coral animals live in a mutualistic relationship with the algae that live inside them. Humans benefit from the mutualistic relationship we have with the *Escherichia coli* that live inside of us. We provide the bacteria with food and comfortable growing conditions, and they help us digest our food and provide us with nutrients.

Interactions of
Living Things

Pre-Reading Questions

Students may not know the answers to these questions before reading the chapter, so accept any reasonable response.

Suggested Answers

1. The deer eats the grass, tree leaves, and acorns. Wolves are common predators of deer. When the deer dies, it becomes food for scavengers and is eventually recycled back into the environment by decomposers.

2. the sun

Interactions of
Living Things

Sections

Pre-Reading
Questions

1. Imagine a deer living in a meadow. What does the deer eat? What eats the deer? When the deer dies what happens to its remains?

2. What is the source of energy for plants?

WHAT'S FOR DINNER?

Look at the harsh, frozen environment surrounding this colony of penguins. What do you suppose the penguins eat? Penguin colonies sometimes number more than a million. They survive by swimming underwater to catch and eat krill, a type of tiny shellfish. Penguins compete with whales—and each other—for the supply of krill. In this chapter, you will learn how living things interact and how energy, in the form of food, travels in pathways through different organism groups.

2

internet connect

 HRW On-line Resources

go.hrw.com
For worksheets and other teaching aids, visit the HRW Web site and type in the keyword: **HSTINT**

 SCiLINKS NSTA

www.scilinks.com
Use the sciLINKS numbers at the end of each chapter for additional resources on the **NSTA** Web site.

 Smithsonian Institution

www.si.edu/hrw
Visit the Smithsonian Institution Web site for related on-line resources.

 CNNfyi.com

www.cnnfyi.com
Visit the CNN Web site for current events coverage and classroom resources.

WHO EATS WHOM?

In this activity, you will learn how organisms interact when finding (or becoming) the next meal.

Procedure

1. On each of four **index cards,** print the name of one of the following organisms: white-tailed deer, turkey vulture, oak tree, and cougar

2. Arrange the cards on your desk in a chain to show who eats whom.

3. List the order of your cards in your ScienceLog.

4. In nature, would you expect to see more cougars, more deer, or more oak trees? Arrange the cards in order of most individuals to fewest.

Analysis

5. What might happen to the other organisms if the oak trees were removed from this group? What might happen if the cougars were removed?

6. Are there any organisms in this group that eat more than one kind of food? (Hint: What else might a deer, a cougar, or a turkey vulture eat?) How could you change the order of your cards to show this information? How could you use pieces of string to show these relationships?

3

START-UP
Activity

WHO EATS WHOM?

MATERIALS

FOR EACH GROUP:
• four 3 x 5 index cards

Answers to START-UP Activity

5. The other organisms might starve if the oak trees were removed. In this arrangement, the deer's only possible foods are oak tree leaves and acorns. And without the deer, the cougar and vulture would starve.

6. The deer, the cougar, and the vulture regularly eat more than one kind of food. But only the vulture would eat more than one of the organisms in this grouping: deer and cougar. Accept any arrangement of cards and string that indicates that deer eat oak, cougars eat deer, and vultures eat deer and cougars. Your best students may arrange the cards in trophic levels.

Focus

Everything Is Connected

In this section, students will learn that ecology is the study of interactions between organisms and their environment. Students will also learn to identify the biotic and abiotic parts of an environment and how environments are organized into populations, communities, and ecosystems.

🔔 Bellringer

Use the board or an overhead projector to display these two sentences as students enter:

> People are a part of nature.
> People are not a part of nature.

Ask students to consider the meanings of these statements and explain them in their ScienceLog.

1) Motivate

DISCUSSION

The *fynbos* is a region in South Africa that contains more than 8,500 native plants, two-thirds of which live nowhere else. Non-native plants are now threatening the survival of the fynbos by outcompeting the local plants for resources such as light, space, and water. Research has shown that there is a connection between the health of the fynbos and the well-being of the people nearby. Ask students the following question:

> Should the non-native plants be removed from the fynbos? That would be very expensive and difficult. What other options are there?

Terms to Learn

ecology   community
biotic   ecosystem
abiotic   biosphere
population

What You'll Do

◆ Distinguish between the biotic and abiotic environment.
◆ Explain how populations, communities, ecosystems, and the biosphere are related.
◆ Explain how the abiotic environment relates to communities.

Figure 1 *The alligator affects, and is affected by, many organisms in its environment.*

Teaching Transparency 65
"The Five Levels of Environmental Organization"

Directed Reading Worksheet Section 1

Everything Is Connected

Look at **Figure 1** below. An alligator drifts in a weedy Florida river, watching a long, thin fish called a gar. The gar swims too close to the alligator. Suddenly, in a rush of snapping jaws and splashing water, the gar becomes a meal for the alligator.

It is clear that these two organisms have just interacted with one another. But organisms have many interactions other than simply "who eats whom." For example, alligators dig underwater holes to escape from the heat. Later, after the alligators abandon these holes, fish and other aquatic organisms live in them when the water level gets low during a drought. Alligators also build nest mounds in which to lay their eggs, and they enlarge these mounds each year. Eventually, the mounds become small islands where trees and other plants grow. Herons, egrets, and other birds build their nests in the trees. It is easy to see that alligators affect many organisms, not just the gars that they eat.

Studying the Web of Life

All living things are connected in a web of life. Scientists who study the connections among living things specialize in the science of ecology. **Ecology** is the study of the interactions between organisms and their environment.

An Environment Has Two Parts An organism's environment is anything that affects the organism. An environment consists of two parts. The **biotic** part of the environment is all of the organisms that live together and interact with one another. The **abiotic** part of the environment includes all of the physical factors—such as water, soil, light, and temperature—that affect organisms living in a particular area. Take another look at Figure 1. How many biotic parts and abiotic parts can you see?

IS THAT A FACT!

Alligators often dig deep holes in the ground. These holes fill up with water and remain full, even during dry winters. Other animals rely heavily on these holes when fresh water is scarce. In Florida, cattle ranchers have used these water sources for their livestock during droughts.

Organization in the Environment At first glance, the environment may seem disorganized. To ecologists, however, the environment can be arranged into different levels, as shown in **Figure 2.** The first level contains the individual organism. The second level contains similar organisms, forming a population. The third contains different populations, forming a community. The fourth contains a community and its abiotic environment, forming an ecosystem. Finally, the fifth level contains all ecosystems, forming the biosphere. Turn the page and examine **Figure 3** to see these levels in a salt marsh.

Figure 2 The Five Levels of Environmental Organization

Organism

Population

Community

Ecosystem

Biosphere

QuickLab

The Human Population

1. Using a **sheet of graph paper,** a **pencil,** and a **ruler,** draw and label a graph as shown below.

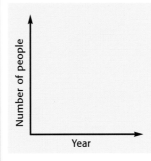

Number of people

Year

2. Plot the following points on your graph:
   (1800, 1 billion people)
   (1930, 2 billion people)
   (1960, 3 billion people)
   (1975, 4 billion people)
   (1987, 5 billion people)
   (1999, 6 billion people)
3. Draw a line connecting the points.
4. Answer the following questions in your ScienceLog.
   a. What does the curve that you have drawn indicate about human population growth?
   b. Do you think the human population can continue to grow indefinitely? Why or why not? *TRY at HOME*

5

Multicultural CONNECTION

The Yanomami are a tribe living in remote rain-forest jungles of Venezuela and Brazil. The Yanomami in Venezuela have had very little contact with outsiders and are one of the few cultures remaining in the world in which people are still an integral part of an intact natural ecosystem. The Yanomami and their land in Venezuela are protected as an international biosphere reserve. Outsiders must get written permission to visit, but even these few visits have given the Yanomami a taste of the outside world. Many observers question how long the Yanomami hunting and gathering culture can last.

Multicultural
CONNECTION

In Hawaii, officials of the Ethnobotanical Garden are working with scientists to develop wetland habitats that will support makaloa, a wetland sedge once used by Hawaiians to weave mats. The wetlands will be part of a waste-water treatment system and will support the resurgence of makaloa. Encourage interested students to research the weaving traditions of Hawaii and to present their findings to the class.

RETEACHING

Writing Ask students to list biotic and abiotic factors in their neighborhood. Write these factors on the board. Then ask students to describe how biotic factors change when abiotic factors change. (For example, if it rains a lot, plants grow more. If there is a drought, plants wither and leaves turn brown.) **Sheltered English**

internetconnect

SCLINKS
NSTA

TOPIC: Organization in the Environment
GO TO: www.scilinks.org
*sci*LINKS NUMBER: HSTL435

Populations A salt marsh is a coastal area where grasslike plants grow. A **population** is a group of individuals of the same species that live together in the same area at the same time. For example, all of the seaside sparrows that live together in a salt marsh are members of a population. The individuals in the population compete with one another for food, nesting space, and mates.

Communities A **community** consists of all of the populations of different species that live and interact in an area. The various animals and plants you see below form a salt-marsh community. The different populations in a community depend on each other for food, shelter, and many other things.

Ecosystems An **ecosystem** is made up of a community of organisms and its abiotic environment. An ecologist studying the salt-marsh ecosystem would examine how the ecosystem's organisms interact with each other and how temperature, precipitation, and soil characteristics affect the organisms. For example, the rivers and streams that empty into the salt marsh carry nutrients, such as nitrogen, from the land. These nutrients influence how the cordgrass and algae grow.

Figure 3 *Examine the picture of a salt marsh below. See if you can find examples of each level of organization in this environment.*

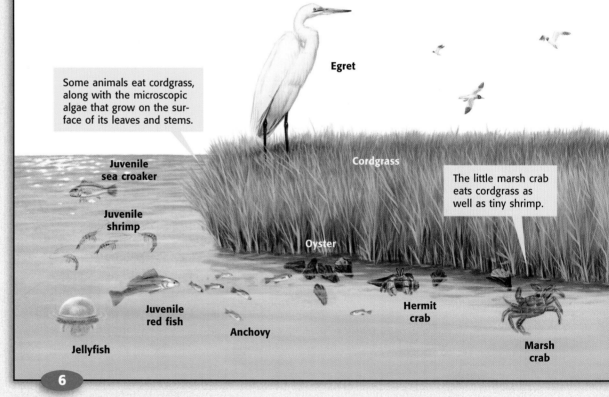

Laughing gull

Some animals eat cordgrass, along with the microscopic algae that grow on the surface of its leaves and stems.

Egret

Cordgrass

The little marsh crab eats cordgrass as well as tiny shrimp.

Juvenile sea croaker

Juvenile shrimp

Oyster

Juvenile red fish

Anchovy

Hermit crab

Jellyfish

Marsh crab

6

CONNECT TO
ENVIRONMENTAL SCIENCE

Wetland communities are more than plant and animal habitats. Wetlands are also natural water-filtration systems that reduce levels of contaminants in municipal waste water. Wetlands also aid in soil formation, the replenishment of ground water, shoreline stabilization, and erosion prevention. In Arizona, California, Hawaii, New Mexico, and Nevada, engineers and biologists have constructed artificial wetlands to study these processes.

The Biosphere The **biosphere** is the part of Earth where life exists. It extends from the deepest parts of the ocean to very high in the atmosphere, where tiny insects and plant spores drift, and it includes every ecosystem. Ecologists study the biosphere to learn how organisms interact with the abiotic environment—Earth's gaseous atmosphere, water, soil, and rock. The water in the abiotic environment includes both fresh water and salt water as well as water that is frozen in polar icecaps and glaciers.

SECTION REVIEW

1. What is ecology?

2. Give two examples of biotic and abiotic factors in the salt-marsh ecosystem.

3. Using the salt-marsh example, distinguish between populations, communities, ecosystems, and the biosphere.

4. **Analyzing Relationships** What do you think would happen to the other organisms in the salt-marsh ecosystem if the cordgrass were to suddenly die?

internetconnect

SCiLINKS
NSTA

TOPIC: Biotic and Abiotic Factors, Organization in the Environment
GO TO: www.scilinks.org
sciLINKS NUMBER: HSTL430, HSTL435

Heron

Laughing gull

Seaside sparrows eat insects, spiders, and small crabs. A male and his mate weave a nest out of cordgrass stalks.

The periwinkle snail eats the algae that grows on the cordgrass. The periwinkle snail also uses the cordgrass as a place to hide from predators.

Clapper rail

Seaside sparrow

Periwinkle snail

Diamondback terrapin

Oyster

7

▼ Answers to Section Review

1. Ecology is the study of the interactions of organisms with each other and with their environment.

2. Biotic factors include all living organisms in the salt marsh. Abiotic factors include all nonliving things in the salt marsh.

3. A population of seaside sparrows lives in a community, which includes all the populations around it. This community lives in a salt-marsh ecosystem, which is defined by a certain geographical location and a certain climate. This ecosystem exists within the biosphere, which is the part of Earth where life exists, including our atmosphere.

4. All the other organisms depend directly or indirectly on cordgrass for food. Without cordgrass, all of the other organisms would probably die out or move into other areas.

---

3 Extend

GOING FURTHER

Writing Have students write a brief report about the effects of temperature on alligator eggs; about the effects of rain, or the lack of rain, on the animals that live in Serengeti National Park, in Tanzania, and the Masai Mara, in Kenya; or about the salinity of the Dead Sea and its effects on organisms that live there.

4 Close

Quiz

1. A caterpillar, a deer, and a rabbit all want to drink from the same puddle, eat the same plant, and bask in the same spot of sunshine. Are they competing members of a population? Why or why not? (No; each is a different species. Therefore, they are competing members of a community.)

2. Using the salt marsh example, explain why ecologists state that saving a large animal, such as the heron or egret, can also save an ecosystem? (The heron and the egret can survive only if the salt marsh remains intact to provide shelter and food for the organisms that the birds need to eat.)

ALTERNATIVE ASSESSMENT

Have students find and draw pictures to create a poster that shows at least five biotic and five abiotic factors of either an aquatic or a terrestrial ecosystem. **Sheltered English**

Focus

Living Things Need Energy

In this section students will learn how producers, consumers, and decomposers obtain energy to survive. They will also learn the difference between a food chain and a food web and how energy is transferred among members of a food chain or a food web. Finally, students will learn to define *habitat* and *niche*.

🔔 Bellringer

Inform students that a plant called Indian pipe is completely white—it has no chlorophyll or chloroplasts. Can this plant still be a producer? If not, where does it get the energy it needs to survive? Have students write their answer in their ScienceLog. (This plant is a consumer. It lives off the roots of rotting trees with the help of a fungus.)

1 Motivate

ACTIVITY

Writing Before reading this section, students should define *producer* and *consumer* in their ScienceLog, using prior knowledge. They should also define *carnivore*, *omnivore*, and *herbivore* in their own words. After reading the section, students should revise their definitions.

internet**connect**

SCI**LINKS** NSTA

TOPIC: Producers, Consumers, and Decomposers
GO TO: www.scilinks.org
*sci***LINKS NUMBER:** HSTL440

SECTION 2
READING WARM-UP

Terms to Learn

herbivore	food web
carnivore	energy pyramid
omnivore	habitat
scavenger	niche
food chain	

What You'll Do

◆ Describe the functions of producers, consumers, and decomposers in an ecosystem.
◆ Distinguish between a food chain and a food web.
◆ Explain how energy flows through a food web.
◆ Distinguish between an organism's habitat and its niche.

Living Things Need Energy

All living things need energy to survive. For example, black-tailed prairie dogs, which live in the grasslands of North America, eat grass and seeds to get the energy they need. They use this energy to grow, move, heal injuries, and reproduce. In fact, everything a prairie dog does requires energy. The same is true for the plants that grow in the grasslands where the prairie dogs live. Coyotes that stalk prairie dogs, as well as the bacteria and fungi that live in the soil, all need energy.

The Energy Connection

Organisms in a prairie or any community can be divided into three groups based on how they obtain energy. These groups are producers, consumers, and decomposers. Examine **Figure 4** to see how energy passes through these groups in an ecosystem.

Producers Organisms that use sunlight directly to make food are called *producers*. They do this using a process called photosynthesis. Most producers are plants, but algae and some bacteria are also producers. Grasses are the main producers in a prairie ecosystem. Examples of producers in other ecosystems include cordgrass and algae in a salt marsh and trees in a forest. Algae are the main producers in the ocean.

Figure 4 *Follow the pathway of energy as it moves from the sun through the ecosystem.*

Energy Sunlight is the source of energy for almost all living things.

Producer Plants use the energy in sunlight to make food.

Consumer The black-tailed prairie dog eats seeds and grass in the grasslands of western North America.

Consumer All of the prairie dogs in a colony watch for enemies, such as coyotes, hawks, and badgers. Occasionally, a prairie dog is killed and eaten by a coyote.

Herbivore

Carnivore

CONNECT TO EARTH SCIENCE

Fossil fuels store solar energy gathered by ancient plants. Use the following to illustrate the formation of coal.

Teaching Transparency 116 "Formation of Coal"

Consumers Organisms that eat producers or other organisms for energy are called *consumers*. They cannot use the sun's energy directly like producers can. Instead, consumers must eat producers or other animals to obtain energy. There are several kinds of consumers. A **herbivore** is a consumer that eats plants. Herbivores in the prairie ecosystem include grasshoppers, gophers, prairie dogs, bison, and pronghorn antelope. A **carnivore** is a consumer that eats animals. Carnivores in the prairie ecosystem include coyotes, hawks, badgers, and owls. Consumers known as **omnivores** eat a variety of organisms, both plants and animals. The grasshopper mouse is an example of an omnivore in the prairie ecosystem. It eats insects, scorpions, lizards, and grass seeds. **Scavengers** are animals that feed on the bodies of dead animals. The turkey vulture is a scavenger in the prairie ecosystem. Examples of scavengers in aquatic ecosystems include crayfish, snails, clams, worms, and crabs.

Decomposers Organisms that get energy by breaking down the remains of dead organisms are called *decomposers*. Bacteria and fungi are examples of decomposers. These organisms extract the last bit of energy from dead organisms and produce simpler materials, such as water and carbon dioxide. These materials can then be reused by plants and other living things. Decomposers are an essential part of any ecosystem because they are nature's recyclers.

BRAIN FOOD

Prairie dogs are not really dogs. They are rodents. They are called dogs because their warning calls sound like the barking of dogs.

Self-Check

Are you a herbivore, a carnivore, or an omnivore? Explain. *(See page 168 to check your answer.)*

Consumer
A turkey vulture may eat some of the coyote's leftovers. A scavenger can pick bones completely clean.

Decomposer
Any prairie dog remains not eaten by the coyote or the turkey vulture are broken down by bacteria and fungi that live in the soil.

Scavenger

Recycler

9

Turkey vultures have an acute sense of smell. A biologist once put decaying carcasses in metal containers, hid the containers in the California foothills, and used a fan to diffuse the odor. Turkey vultures were soon soaring overhead.

Engineers once pumped ethyl mercaptan, which smells like rotting flesh, into natural-gas lines. They located leaks by watching for turkey vultures attracted to the pipeline.

Answer to Self-Check

A food chain shows how energy moves in one direction from one organism to the next. A food web shows that there are many energy pathways between organisms.

MATH and MORE

There are 12,000 units of the sun's energy available to grass at the base of an energy pyramid. Grass stores in its tissues 10 percent of the available energy, so that energy becomes available to the next consumer, a rabbit. The rabbit, a consumer of grass, stores 10 percent of the energy that was stored in the grass. A coyote, a consumer of rabbits, stores 10 percent of the energy that was stored in the rabbit. Calculate the units of food energy stored in the grass, the rabbit, and the coyote.

The grass stores 10 percent of the sun's energy:

$0.1 \times 12{,}000 = 1{,}200$ units of energy stored in the grass

The rabbit stores 10 percent of the grass' stored energy:

$0.1 \times 1{,}200 = 120$ units of energy stored in the rabbit

The coyote stores 10 percent of the rabbit's stored energy:

$0.1 \times 120 = 12$ units of energy stored in the coyote

Math Skills Worksheet
"Working with Percentages and Proportions"

Self-Check

How is a food web different from a food chain? *(See page 168 to check your answer.)*

Figure 5 *Energy moves through an ecosystem in complex ways. Most consumers eat a variety of foods and can be eaten by a variety of other consumers.*

Food Chains and Food Webs

Figure 4, on pages 8–9, shows a **food chain,** which represents how the energy in food molecules flows from one organism to the next. But because few organisms eat just one kind of organism, simple food chains rarely occur in nature. The many energy pathways possible are more accurately shown by a **food web.** **Figure 5** shows a simple food web for a woodland ecosystem.

Find the fox and the rabbit in the figure below. Notice that the arrow goes from the rabbit to the fox, showing that the rabbit is food for the fox. The rabbit is also food for the owl. Neither the fox nor the owl is ever food for the rabbit. Energy moves from one organism to the next in a one-way direction, even in a food web. Any energy not immediately used by an organism is stored in its tissues. Only the energy stored in an organism's tissues can be used by the next consumer.

internet connect

SCI LINKS
NSTA

TOPIC: Food Chains and Food Webs
GO TO: www.scilinks.org
*sci*LINKS NUMBER: HSTL445

SCIENCE HUMOR

Customer: Waiter! Waiter! There's a fly in my soup!

Waiter: Don't worry, sir, the spider in your salad will get it!

Energy Pyramids

A grass plant uses most of the energy it obtains from the sun for its own life processes. But some of the energy is stored in its tissues and is left over for prairie dogs and other animals that eat the grass. Prairie dogs need a lot of energy and have to eat a lot of grass. Each prairie dog uses most of the energy it obtains from eating grass and stores only a little of it in its tissues. Coyotes need even more energy than prairie dogs, so they must eat many prairie dogs to survive. There must be many more prairie dogs in the community than there are coyotes that eat prairie dogs.

The loss of energy at each level of the food chain can be represented by an **energy pyramid,** as shown in **Figure 6.** You can see that the energy pyramid has a large base and becomes smaller at the top. The amount of available energy is reduced at higher levels because most of the energy is either used by the organism or given off as heat. Only energy stored in the tissues of an organism can be transferred to the next level.

MATH BREAK

Energy Pyramids

Draw an energy pyramid for a river ecosystem that contains four levels—aquatic plants, insect larvae, bluegill fish, and a largemouth bass. The plants obtain 10,000 units of energy from the sun. If each level uses 90 percent of the energy it receives from the previous level, how many units of energy are available to the bass?

ACTIVITY

Writing **Inferring Information**
Chipmunks, tree squirrels, and flying squirrels are all related animals that eat seeds. Yet each has its own niche, or way of life, that serves to limit the amount of competition between the species.

Have students describe in writing, to the best of their knowledge, the niche, or way of life, for each of these seed eaters.

(**Chipmunks** live underground, collect small seeds, and store these seeds in cheek pouches until depositing them in an underground storage area.

Tree squirrels live in trees, collect larger seeds and nuts one at a time, and bury them in the ground.

Flying squirrels are active only at night; live in trees; eat large and small seeds, fruits, and insects; and glide from tree to tree using flaps of skin.)

Teaching Transparency 66 "Energy Pyramid"

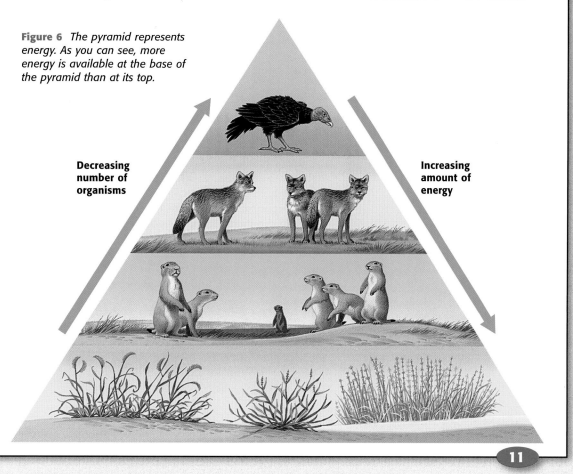

Figure 6 *The pyramid represents energy. As you can see, more energy is available at the base of the pyramid than at its top.*

Decreasing number of organisms

Increasing amount of energy

11

IS THAT A FACT!

In 1989, the Nature Conservancy purchased 30,000 acres of grassland in Oklahoma. The conservancy's goal is "the restoration of a functioning tall-grass prairie ecosystem." The land has been grazed by cattle but never plowed; the restoration will allow the more than 700 prairie plant species to reestablish themselves. A healthy prairie is also home to 300 bird species, 80 mammal species, and hundreds of thousands of insect species. Biologists have reintroduced bison, whose grazing is an integral part of the prairie food web.

Figure 7 *As the wilderness was settled, the gray wolf population in the United States declined.*

Figure 8 *Members of the U.S. Fish and Wildlife Service are moving a caged wolf to a location in Yellowstone National Park.*

Wolves and the Energy Pyramid

A single species can be very important to the flow of energy in an environment. Gray wolves, for example, are a consumer species that can control the populations of many other species. The diet of gray wolves can include anything from a lizard to an elk.

Once common throughout much of the United States, gray wolves were almost wiped out as the wilderness was settled. You can see a pair of gray wolves in **Figure 7.** Without wolves, certain other species, such as elk, were no longer controlled. The overpopulation of elk in some areas led to overgrazing and starvation.

Gray wolves were recently restored to the United States at Yellowstone National Park, as shown in **Figure 8.** The U.S. Fish and Wildlife Service hopes this action will restore the natural energy flow in this wilderness area. Not everyone approves, however. Ranchers near Yellowstone are concerned about the safety of their livestock.

Habitat and Niche

An organism's **habitat** is the environment in which it lives. The wolf's habitat was originally very extensive. It included forests, grasslands, deserts, and the northern tundra. Today the wolf's habitat in North America is much smaller. It includes wilderness areas in Montana, Washington, Minnesota, Michigan, Wisconsin, and Canada.

An organism's way of life within an ecosystem is its **niche.** An organism's niche includes its habitat, its food, its predators, and the organisms with which it competes. An organism's niche also includes how the organism affects and is affected by abiotic factors in its environment, such as temperature, light, and moisture.

The Niche of the Gray Wolf

A complete description of a species' niche is very complex. To help you distinguish between habitat and niche, parts of the gray wolf niche are described on the next page.

12

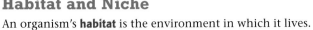

WEIRD SCIENCE

In certain tropical areas, the dry season can last several months. Some frogs, such as the African bullfrog, burrow into the mud before the mud dries out. Then the frog sheds several intact layers of skin, which harden and form a protective covering. Only the frog's nostrils are exposed, so it can breathe. When the rains return, the frog breaks through the moist soil and returns to the surface.

Gray Wolves Are Consumers Wolves are carnivores. Their diet includes large animals, such as deer, moose, reindeer, sheep, and elk, as well as small animals, such as birds, lizards, snakes, and fish.

Gray Wolves Have a Social Structure Wolves live and hunt in packs, which are groups of about six animals that are usually members of the same family. Each member of the pack has a particular rank within the pack. The pack has two leaders that help defend the pack against enemies, such as other wolf packs or bears.

Gray Wolves Nurture and Teach Their Young A female wolf, shown in **Figure 9,** has five to seven pups and nurses her babies for about 2 months. The entire pack help bring the pups food and baby-sit when the parents are away from the den. It takes about 2 years for the young wolves to learn to hunt. At that time, some young wolves leave the pack to find mates and start their own pack.

Figure 9 *In small wolf packs, only the alpha female has pups. They are well cared for, however, by all of the males and females in the pack.*

Gray Wolves Are Needed in the Food Web If wolves become reestablished at Yellowstone National Park, they will reduce the elk population by killing the old, injured, and diseased elk. This in turn will allow more plants to grow, which will allow animals that eat the plants, such as snowshoe hares, and the animals that eat the hares, such as foxes, to increase in number.

SECTION REVIEW

1. How are producers, consumers (herbivores, carnivores, and scavengers), and decomposers linked in a food chain?

2. How do food chains link together to form a food web?

3. Distinguish between an organism's habitat and its niche using the prairie dog as an example.

4. **Applying Concepts** Is it possible for an inverted energy pyramid to exist, as shown in the figure at right? Explain why or why not.

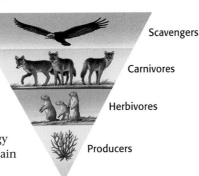

Scavengers

Carnivores

Herbivores

Producers

13

▼ **Answers to Section Review**

1. Producers convert sunlight into food. Herbivores and carnivores are both consumers. Herbivores eat producers, and carnivores eat herbivores and other consumers. Scavengers eat the bodies of dead animals. Decomposers break down the remains of dead animals and extract the last bit of energy from them.

2. Food chains show how energy is passed through an ecosystem from producers to decomposers. A food web shows how all the food chains of an ecosystem are connected.

3. The prairie is a prairie dog's habitat. What it affects and what affects it make up its niche.

4. No; the widest part of the pyramid has the most energy, and the pointed end has the least. An ecosystem cannot support the greatest energy level with the level that has the least energy.

Focus

Types of Interactions

In this section, students will learn about the types of interactions that organisms have with each other and with their environment. Students will learn to distinguish the two types of competition and to identify predator and prey species. They will also learn to recognize mutualism, commensalism, and parasitism. Finally, students will learn to define *coevolution*.

Bellringer

On the board or overhead projector, write the following:

In your ScienceLog, make a list of predators that are also prey. (Answers may include salamanders, frogs, shrews, snakes, lizards, and weasels.) **Sheltered English**

1) Motivate

DISCUSSION

In India, the chital, a small deer, has trouble finding enough grass to eat during the dry season. The deer rely on help from a type of monkey to get enough to eat. Ask students:

Can you guess how the monkeys help the deer? (This particular monkey is a messy and finicky eater, dropping leaves as it feeds in the trees. The deer then gobble up the food.)

Which species benefit from this relationship? (It seems that only the deer benefit.)

Now tell students that the deer have keen eyesight, hearing, and sense of smell. They therefore help warn the monkeys of predators. Now which species do you think benefit? (Both species benefit.)

Terms to Learn

carrying capacity	mutualism
prey	commensalism
predator	parasitism
symbiosis	coevolution

What You'll Do

- Distinguish between the two types of competition.
- Give examples of predators and prey.
- Distinguish between mutualism, commensalism, and parasitism.
- Define *coevolution*, and give an example.

Types of Interactions

Look at the seaweed forest shown in **Figure 10** below. Notice that some types of organisms are more numerous than others. In natural communities, populations of different organisms vary greatly. The interactions between these populations affect the size of each population.

Figure 10 *This seaweed forest is home to a large number of interacting species.*

Interactions with the Environment

Most living things produce more offspring than will survive. A female frog, for example, might lay hundreds of eggs in a small pond. In a few months, the population of frogs in that pond will be about the same as it was the year before. Why won't the pond become overrun with frogs? An organism, such as a frog, interacts with biotic or abiotic factors in its environment that can control the size of its population.

Limiting Factors Populations cannot grow indefinitely because the environment contains only so much food, water, living space, and other needed resources. When one or more of those resources becomes scarce, it is said to be a *limiting factor*. For example, food becomes a limiting factor when a population becomes too large for the amount of food available. Any single resource can be a limiting factor to population size.

Green herons make interesting use of the biotic and abiotic parts of their environment in Japan. They will drop sticks and even bread crumbs into the water to attract fish. Sometimes they catch a fish just 2–3 seconds after they drop the bait.

Carrying Capacity The largest population that a given environment can support over a long period of time is known as the environment's **carrying capacity.** When a population grows larger than its carrying capacity, limiting factors in the environment cause the population to get smaller. For example, after a very rainy growing season in an environment, plants may produce a large crop of leaves and seeds. This may cause a herbivore population to grow large because of the unlimited food supply. If the next year has less rainfall than usual, there won't be enough food to support the large herbivore population. In this way, a population may temporarily exceed the carrying capacity. But a limiting factor will cause the population to die back. The population will return to a size that the environment can support over a long period of time.

Interactions Among Organisms

Populations contain interacting individuals of a single species, such as a group of rabbits feeding in the same area. Communities contain interacting populations of several species, such as a coral reef community with many species trying to find living space. Ecologists have described four main ways that species and individuals affect each other: competition, predators and prey, certain symbiotic relationships, and coevolution.

Competition

When two or more individuals or populations try to use the same limited resource, such as food, water, shelter, space, or sunlight, it is called *competition*. Because resources are in limited supply in the environment, their use by one individual or population decreases the amount available to other organisms.

Competition can occur among individuals *within* a population. The elks in Yellowstone National Park are herbivores that compete with each other for the same food plants in the park. This is a big problem for this species in winter. Competition can also occur *between* populations of different species. The different species of trees in **Figure 11** are competing with each other for sunlight and space.

Figure 11 *Some of the trees in this forest grow tall in order to reach sunlight, reducing the amount of sunlight available to shorter trees nearby.*

15

Answers to Self-Check

1. If an area has only enough water to support 10 organisms, any additional organisms will cause some to go without water and move away or die.
2. Weather favorable for growing the food the deer eat will allow the forest to support more deer.

DISCUSSION

Examining Cycles The lynx relies primarily upon the snowshoe hare for food. Ask students:

How will an abundance of plant food affect the hare population? (The population will increase.)

What will then happen to the lynx population? (It will also increase.)

What will happen when the hares begin to run out of food? (Hares will grow weak from hunger and will be easily caught by lynxes. The lynx population will grow until so many hares are killed that the lynxes also begin to die off.)

How can the hare and lynx populations recover? (The plants must recover from the overbrowsing.)

Explain that observers have documented 10-year cycles of peaks and crashes in the lynx and snowshoe hare populations for the past 200 years. Studies have shown that snowshoe hares have the same population cycles when no lynxes are present. Lynx population cycles, however, are dependent on the availability of hares.

MISCONCEPTION ///ALERT\\\

While it's true that all members of an ecosystem have important roles, some members are more important than others to the overall health of the ecosystem. Such species are called keystone species. Like removing the keystone that holds up the other stones in an arch, removing a keystone species can cause the whole system to collapse. This occurred when the sea otter was hunted to near extinction in the kelp forests of the Pacific Ocean. With the otters gone, nothing was left to eat the sea urchins, which multiplied and ate all the kelp. The kelp was home to dozens of animals, all of which disappeared, along with most of the sea urchins. When the otter was reintroduced, all these organisms returned!

Directed Reading Worksheet Section 3

COOPERATIVE LEARNING

Divide students into six groups. Then divide each group into subgroups of two or three, and designate each subgroup as Predator or Prey. Instruct each Predator group to name a prey animal and the adaptation it uses to catch the prey. (speed, talons, strength)

Tell each Prey group to name a predator animal and its adaptations to evade capture. (camouflage, speed, protective armor, flight)

Have students write down the adaptations and share their information with the other groups. Sheltered English

BRAIN FOOD

Studies have shown that ducks can have one-half of their brain asleep while the other half is awake. The side that is awake can even have an eye open and respond to stimuli. The sides can also switch so that the half that was awake can rest and the rested half can become alert. Ask students why this ability might be helpful to ducks. (Ducks need sleep, but when they sleep they are vulnerable to predators. Having the ability to keep an eye open and be alert to danger while part of their brain rests helps them avoid predators. Because they live in flocks, they can arrange themselves in a resting spot so that the ducks on the outside edge of the flock each have an open eye looking in a different direction.)

Predators and Prey

Many interactions among species occur because one organism eats another. The organism that is eaten is called the **prey.** The organism that eats the prey is called the **predator.** When a bird eats a worm, the worm is the prey and the bird is the predator.

Predator Adaptations In order to survive, predators must be able to catch their prey. Predators have a wide variety of methods and abilities for doing this. The cheetah, for example, is able to run at great speed to catch its prey. Other predators, such as the goldenrod spider, shown in **Figure 12,** ambush their prey. The goldenrod spider blends in so well with the goldenrod flower that all it has to do is wait for its next insect meal to arrive.

Figure 12 The goldenrod spider is difficult for its insect prey to see. Can you see it?

Prey Adaptations Prey organisms have their own methods and abilities to keep from being eaten. Prey are able to run away, stay in groups, or camouflage themselves. Some prey organisms are poisonous to predators. They may advertise their poison with bright colors to warn predators to stay away. The fire salamander, shown in **Figure 13,** sprays a poison that burns. Predators quickly learn to recognize its warning coloration.

Many animals run away from predators. Prairie dogs run to their underground burrows when a predator approaches. Many small fishes, such as anchovies, swim in groups called schools. Antelopes and buffaloes stay in herds. All the eyes, ears, and noses of the individuals in the group are watching, listening, and smelling for predators. This behavior increases the likelihood of spotting a potential predator.

Some prey species hide from predators by using camouflage. Certain insects resemble leaves so closely that you would never guess they are animals.

Figure 13 Experienced predators know better than to eat the fire salamander! This colorful animal will make an unlucky predator very sick.

MISCONCEPTION ///ALERT\\\

The phrase *balance of nature* does not imply that the components of an environment are static. Populations are in balance when their sizes are stable. That stability is the result of constant interactions between individuals of a population, between populations in a community, and between populations and environmental factors.

WEIRD SCIENCE

Pancake tortoises live on rocky hillsides in Africa. They are very flat and can wedge themselves into cracks in the rocks for protection from predators. Their bottom shells are pliable and can "inflate" so that a predator can't pry the tortoises out.

Symbiosis

Some species have very close interactions with other species. **Symbiosis** is a close, long-term association between two or more species. The individuals in a symbiotic relationship can benefit from, be unaffected by, or be harmed by the relationship. Often, one species lives in or on the other species. The thousands of symbiotic relationships that occur in nature are often classified into three groups: mutualism, commensalism, and parasitism.

Mutualism A symbiotic relationship in which both organisms benefit is called **mutualism.** For example, you and a species of bacteria that lives in your intestines benefit each other! The bacteria get a plentiful food supply from you, and in return you get vitamins that the bacteria produce.

Another example of mutualism occurs between coral and algae. The living corals near the surface of the water provide a home for the algae. The algae produce food through photosynthesis that is used by the corals. When a coral dies, its skeleton serves as a foundation for other corals. Over a long period of time, these skeletons build up large, rocklike formations that lie just beneath the surface of warm, sunny seas, as shown in **Figure 14.**

Figure 14 *In the smaller photo above, you can see the gold-colored algae inside the coral.*

Commensalism A symbiotic relationship in which one organism benefits and the other is unaffected is called **commensalism.** One example of commensalism is the relationship between sharks and remoras. **Figure 15** shows a shark with a remora attached to its body. Remoras "hitch a ride" and feed on scraps of food left by sharks. The remoras benefit from this relationship, while sharks are unaffected.

Figure 15 *The remora attached to the shark benefits from the relationship. The shark is neither benefited nor harmed.*

Q: What is a parasite's favorite party?

A: a louse-warming party

17

Learners Having Difficulty

Reinforce the concept of symbiosis by presenting the following examples and asking students to categorize the relationship:

1. Acacia ants live on the bullhorn acacia tree, which provides the ants' food and shelter. The ants deter browsing animals who want to eat the tree. (mutualism)

2. Plants called epiphytes, such as lianas and certain orchids, live on other plants, which provide only a substrate. Epiphytes absorb sunlight, water, and nutrients from their surroundings and make their own food. (commensalism)

3. There is a tiny wasp that lays its eggs in a variety of insects, such as caterpillars, spiders, aphids, and flies. The wasp larvae feed on the host insect, eventually killing it. (parasitism)

Sheltered English

DEBATE

Bird Behavior The brown-headed cowbird is a parasite. It lays its eggs in the nests of other birds, which incubate the cowbird's eggs and raise the chicks. Some of the host species are smaller than the cowbird. Their own chicks cannot compete with the cowbird chicks for food and sometimes starve. Some people view the cowbird as a pest that harms other bird species. Others note that it has an adaptation that ensures its survival, no different from any other competitive advantage. Have students discuss these two points of view.

Parasitism A symbiotic association in which one organism benefits while the other is harmed is called **parasitism.** The organism that benefits is called the *parasite.* The organism that is harmed is called the *host.* The parasite gets nourishment from its host, which is weakened in the process. Sometimes a host organism becomes so weak that it dies. Some parasites, such as ticks, live outside the host's body. Other parasites, such as tapeworms, live inside the host's body.

Figure 16 shows a bright green caterpillar called a tomato hornworm. A female wasp laid tiny eggs on the caterpillar. When the eggs hatch, each young wasp will burrow into the caterpillar's body. The young wasps will actually eat the caterpillar alive! In a short time, the caterpillar will be almost completely consumed and will die. When that occurs, the mature wasps will fly away.

In this example of parasitism, the host dies. Most parasites, however, do not kill their hosts. Can you think of reasons why?

Figure 16 *The tomato hornworm is being parasitized by young wasps. Do you see their cocoons?*

Coevolution

Symbiotic relationships and other interactions among organisms in an ecosystem may cause coevolution. **Coevolution** is a long-term change that takes place in two species because of their close interactions with one another.

Coevolution sometimes occurs between herbivores and the plants on which they feed. For example, the ants shown in **Figure 17** have coevolved with a tropical tree called the acacia. The ants protect the tree on which they live by attacking any other herbivore that approaches the tree. The plant has coevolved special structures on its stems that produce food for the ants. The ants live in other structures also made by the tree.

Figure 17 *Ants collect food made by the acacia tree and store the food in their shelter, also made by the tree.*

18

In Africa, billions of dollars a year are spent to destroy a parasitic plant that lives off important crops such as sorghum. The parasite weakens the plant, preventing it from producing its much-needed grains. Recently, a sorghum plant that is resistant to the parasite was developed. Ethiopian farmers were so eager to get started again that they smuggled the new seeds in from Sudan before their own government approved the seeds.

APPLY

Coevolution in Australia

In 1859, settlers released 12 rabbits in Australia. There were no predators or parasites to control the rabbit population, and there was plenty of food. The rabbit population increased so fast that the country was soon overrun by rabbits. To control the rabbit population, the Australian government introduced a virus that makes rabbits sick. The first time the virus was used, more than 99 percent of the rabbits died. The survivors reproduced, and the

rabbit population grew large again. The second time the virus was used, about 90 percent of the rabbits died. Once again, the rabbit population increased. The third time the virus was used, only about 50 percent of the rabbits died. Suggest what changes might have occurred in the rabbits and the virus.

Coevolution and Flowers Some of the most amazing examples of coevolution are between flowers and their pollinators. (An organism that carries pollen from flower to flower is called a *pollinator.*) When the pollinator travels to the next flower to feed, some of the pollen is left behind on the female part of the flower and more pollen is picked up. Because of pollination, reproduction can take place in the plant. Organisms such as bees, bats, and hummingbirds are attracted to a flower because of its colors, odors, and nectar.

During the course of evolution, hummingbird-pollinated flowers, for example, developed nectar with just the right amount of sugar for their pollinators. The hummingbird's long, thin tongue and beak coevolved to fit into the flowers so that they could reach the nectar. As the hummingbird, like the one shown in **Figure 18,** feeds on the nectar, its head and body become smeared with pollen.

Figure 18 *The bird is attracted to the flower's nectar and picks up the flower's pollen as it feeds.*

SECTION REVIEW

1. Briefly describe one example of a predator-prey relationship. Identify the predator and the prey.

2. Name and define the three kinds of symbiosis.

3. **Analyzing Relationships** Explain the probable relationship between the giant *Rafflesia* flower, which smells like rotting meat, and the carrion flies that buzz around it. HINT: *carrion* means "rotting flesh."

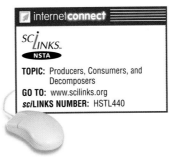

internet connect

SCILINKS.
NSTA

TOPIC: Producers, Consumers, and Decomposers
GO TO: www.scilinks.org
*sci*LINKS NUMBER: HSTL440

19

▼ *Answers to Section Review*

1. Accept any answer in which one animal (the predator) is eating another (the prey).

2. *mutualism*—a symbiotic relationship in which both organisms benefit

   commensalism—a symbiotic relationship in which one organism benefits and the other is unaffected

parasitism—a symbiotic association in which one organism benefits while the other is harmed

3. The probable relationship between *Rafflesia* and carrion flies is a symbiotic mutualism brought about by coevolution.

4) Close

Quiz

1. Explain the difference between mutualism and coevolution. (Mutualism is a close, long-term association between two organisms in which both benefit. Coevolution is the gradual change in two organisms' physical characteristics as a result of a symbiotic relationship.)

2. Can a predator ever be the prey for another species? (Yes; field mice eat insects and are sometimes eaten by snakes and hawks. Small fish are consumed by larger fish, which are eaten by even larger fish.)

ALTERNATIVE ASSESSMENT

Concept Mapping Have students organize the following terms into a concept map:

competition, predator, individuals, population, symbiosis, commensalism, prey, mutualism, interactions

Answer to APPLY

Viruses that kill all of their hosts have no place to live. The viruses in Australia that changed to become less virulent and avoided killing rabbits had more hosts to occupy. The rabbits, on the other hand, began to produce more survivors with immunity. The survivors bred, producing a population of rabbits that included more individuals that were immune to the virus.

 Reinforcement Worksheet "Symbiotic Relationships"

 Interactive Explorations CD-ROM "What's Bugging You?"

Making Models Lab

Adaptation: It's a Way of Life
Teacher's Notes

Time Required

One or two 45-minute class periods

Lab Ratings

⚗	⚗⚗	⚗⚗⚗	⚗⚗⚗⚗
EASY			HARD

TEACHER PREP ⚗⚗
STUDENT SET-UP ⚗⚗
CONCEPT LEVEL ⚗⚗
CLEAN UP ⚗

MATERIALS

The materials listed on the student page are enough for a group of 4–5 students. Materials for this activity can include recycled materials, glue, buttons, pipe cleaners, poster paints and brushes, and any number of other art or craft supplies. Have students bring in as much as they can for their own project and to share.

Lab Notes

This lab provides an opportunity for students to exercise a great deal of creativity and to expand their understanding of adaptations. Most students will probably enjoy choosing the adaptations and inventing a niche where those adaptations are useful. Help them understand that environments and adaptations usually evolve together in the natural world, not one before the other. You may want to make sets of the adaptations and put them in a container for students to draw from. Doing this may lessen duplication in the classroom. Someone might be challenged to design a flying decomposer with armor!

Adaptation: It's a Way of Life

Did you know that organisms have special characteristics called *adaptations* that help them survive changes in their environment? These changes can be climate changes, less food, or disease. These things can cause a population to die out unless some members have adaptations that help them survive. For example, a bird may have an adaptation for eating sunflower seeds and ants. If the ants die out, the bird can still eat seeds in order to live.

In this activity, you will design an organism with special adaptations. Then you will describe how these adaptations help the organism live.

MATERIALS

* poster board
* colored markers
* magazines for cutouts
* other arts-and-crafts materials
* scissors

Procedure

1. Study the chart on the next page. Choose one adaptation from each column. For example, an organism might be a scavenger that burrows underground and has spikes on its tail.

2. Design an organism that has the three adaptations you have chosen. Use poster board, colored markers, picture cutouts, construction paper, or any materials of your choice to create your organism.

3. Write a caption on your poster describing your organism. Describe its appearance, where it lives, and how its adaptations help it survive. Give your animal a two-part "scientific" name based on its characteristics.

4. Display your creation in your classroom. Share with classmates how you chose the adaptations for your organism.

Analysis

5. What does your imaginary organism eat?

6. In what environment would your organism be most likely to survive—in the desert, tropical rain forest, plains, icecaps, mountains, or ocean? Explain your answer.

7. What kind of animal is your organism (mammal, insect, reptile, bird, fish)? What modern organism (on Earth today) or ancient organism (extinct) is your imaginary organism most like? Explain the similarities between the two organisms. Do some research outside of class about a real organism that your imaginary organism may resemble.

20

 Datasheets for LabBook

Alonda Droege
Pioneer Middle School
Steilacom, Washington

8 If a sudden climate change occurred, such as daily down-pours of rain in a desert, would your imaginary organism survive? What adaptations for surviving such a change does it have?

Adaptations		
Diet	**Type of transportation**	**Special adaptation**
• Carnivore • Herbivore • Omnivore • Scavenger • Decomposer	• Flies • Glides through the air • Burrows underground • Runs fast • Swims • Hops • Walks • Climbs • Floats • Slithers	• Uses sensors to detect heat • Is active only at night and has excellent night vision • Changes color to match its surroundings • Has armor • Has horns • Can withstand extreme temperature changes • Secretes a terrible and sickening scent • Has poison glands • Has specialized front teeth • Has tail spikes • Stores oxygen in its cells so it does not have to breathe continuously • One of your own invention

Answers

5–7. All answers will depend on the adaptations that the student chose and the organism the student invented. Students should relate one or more adaptations to the kinds of food the organism eats, where the animal lives, or what kind of animal it is.

8. Several adaptations can be given, but expect a few to say their animals wouldn't survive.

Chapter Highlights

Chapter Highlights

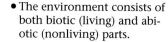

VOCABULARY DEFINITIONS

SECTION 1

ecology the study of the interactions between organisms and their environment

biotic living factors in the environment

abiotic nonliving factors in the environment

population a group of individuals of the same species that live together in the same area at the same time

community all of the populations of different species that live and interact in an area

ecosystem a community of organisms and their nonliving environment

biosphere the part of the Earth where life exists

SECTION 2

herbivore a consumer that eats plants

carnivore a consumer that eats animals

omnivore a consumer that eats a variety of organisms

scavenger an animal that feeds on the bodies of dead animals

food chain a diagram that represents how the energy in food molecules flows from one organism to the next

food web a complex diagram representing the many energy pathways in a real ecosystem

energy pyramid a diagram shaped like a triangle showing the loss of energy at each level of the food chain

habitat the environment where an organism lives

niche an organism's way of life and its relationships with its abiotic and biotic environment

---

SECTION 1

Vocabulary

ecology (p. 4)
biotic (p. 4)
abiotic (p. 4)
population (p. 6)
community (p. 6)
ecosystem (p. 6)
biosphere (p. 7)

Section Notes

- Ecology is the study of the interactions between organisms and their environment.

- The environment consists of both biotic (living) and abiotic (nonliving) parts.

- A population is a group of the same species living in the same place at the same time. A community is all of the populations of different species living together. An ecosystem is a community and its abiotic environment. The biosphere consists of all of Earth's ecosystems.

Labs

Capturing the Wild Bean (p. 128)

SECTION 2

Vocabulary

herbivore (p. 9)
carnivore (p. 9)
omnivore (p. 9)
scavenger (p. 9)
food chain (p. 10)
food web (p. 10)
energy pyramid (p. 11)
habitat (p. 12)
niche (p. 12)

Section Notes

- Organisms that use sunlight directly to make food are called producers. Consumers are organisms that eat other organisms to obtain energy. Decomposers are bacteria and fungi that break down the remains of dead organisms to obtain energy.

☑ Skills Check

Math Concepts

ENERGY PYRAMIDS Try calculating the MathBreak on page 11 as if each unit of energy were $1.00. If you have $10,000.00, but you spend 90 percent, how much do you have left to leave in your will? ($1,000.00) If your heir spends 90 percent of that, how much can your heir leave? ($100.00) After four generations, how much will the inheritance be? ($1.00) Not much, huh? That's why there are very few large organisms at the top of the energy pyramid.

Visual Understanding

FOOD WEBS Several food pathways are shown in the food web in Figure 5 on page 10. However, an actual food web in a woodland ecosystem is much more complex because hundreds of species live there. Find the mouse in Figure 5. How many organisms feed on the mouse? How many organisms feed on the butterfly? What might happen to this ecosystem if these animals were eliminated?

22

Lab and Activity Highlights

Adaptation: It's a Way of Life PG 20

Capturing the Wild Bean PG 128

Datasheets for LabBook (blackline masters for these labs)

SECTION 2

- A food chain shows how energy flows from one organism to the next.

- Because most organisms eat more than one kind of food, there are many energy pathways possible; these are represented by a food web.

- Energy pyramids demonstrate that most of the energy at each level of the food chain is used up at that level and is unavailable for organisms higher on the food chain.

- An organism's habitat is the environment in which it lives. An organism's niche is its role in the ecosystem.

SECTION 3

Vocabulary

carrying capacity *(p. 15)*
prey *(p. 16)*
predator *(p. 16)*
symbiosis *(p. 17)*
mutualism *(p. 17)*
commensalism *(p. 17)*
parasitism *(p. 18)*
coevolution *(p. 18)*

Section Notes

- Population size changes over time.

- Limiting factors slow the growth of a population. The largest population that an environment can support over a long period of time is called the carrying capacity.

- When one organism eats another, the organism that is eaten is the prey, and the organism that eats the prey is the predator.

- Symbiosis is a close, long-term association between two or more species. There are three general types of symbiosis: mutualism, commensalism, and parasitism.

- Coevolution involves the long-term changes that take place in two species because of their close interactions with one another.

SECTION 3

carrying capacity the largest population that a given environment can support over a long period of time

prey an organism that is eaten by another organism

predator an organism that eats other organisms

symbiosis a close, long-term association between two or more species

mutualism a symbiotic relationship in which both organisms benefit

commensalism a symbiotic relationship in which one organism benefits and the other is unaffected

parasitism a symbiotic association in which one organism benefits while the other is harmed

coevolution long-term changes that take place in two species because of their close interactions with one another

  Vocabulary Review Worksheet

Blackline masters of these Chapter Highlights can be found in the **Study Guide.**

---

 internetconnect

 go.hrw.com

GO TO: go.hrw.com

Visit the **HRW** Web site for a variety of learning tools related to this chapter. Just type in the keyword:

KEYWORD: HSTINT

SCILINKS.

N S T A

GO TO: www.scilinks.org

Visit the **National Science Teachers Association** on-line Web site for Internet resources related to this chapter. Just type in the *sci*LINKS number for more information about the topic:

TOPIC: Biotic and Abiotic Factors | *sci*LINKS NUMBER: HSTL430
TOPIC: Organization in the Environment | *sci*LINKS NUMBER: HSTL435
TOPIC: Producers, Consumers, and Decomposers | *sci*LINKS NUMBER: HSTL440
TOPIC: Food Chains and Food Webs | *sci*LINKS NUMBER: HSTL445
TOPIC: Habitats and Niches | *sci*LINKS NUMBER: HSTL450

23

Lab and Activity Highlights

LabBank

 Whiz-Bang Demonstrations, Voracious Fly Catcher

EcoLabs & Field Activities, Survival Is Just a Roll of the Dice

Long-Term Projects & Research Ideas, Out of House and Home

Interactive Explorations CD-ROM

 CD 2, Exploration 1, "What's Bugging You?"

Chapter Review

USING VOCABULARY

To complete the following sentences, choose the correct term from each pair of terms listed below:

1. An organism's environment has two parts, the __?__, or living, and the __?__, or nonliving. *(biotic or abiotic)*

2. A __?__ is a group of individuals of the same species that live in the same area at the same time. *(community or population)*

3. A community and its abiotic environment make up a(n) __?__. *(ecosystem or food web)*

4. Organisms that use photosynthesis to obtain energy are called __?__. *(producers or decomposers)*

5. The environment in which an organism lives is its __?__, and the role the organism plays in an ecosystem is its __?__. *(niche or habitat)*

UNDERSTANDING CONCEPTS

Multiple Choice

6. A tick sucks blood from a dog. In this relationship, the tick is the __?__ and the dog is the __?__.
   a. parasite, prey
   b. predator, host
   c. parasite, host
   d. host, parasite

7. Resources such as water, food, or sunlight are more likely to be limiting factors
   a. when population size is decreasing.
   b. when predators eat their prey.
   c. when the population is small.
   d. when a population is approaching the carrying capacity.

8. "Nature's recyclers" are
   a. predators.
   b. decomposers.
   c. producers.
   d. omnivores.

9. A beneficial association between coral and algae is an example of
   a. commensalism.
   b. parasitism.
   c. mutualism.
   d. predation.

10. How energy moves through an ecosystem can be represented by
    a. food chains.
    b. energy pyramids.
    c. food webs.
    d. All of the above

11. The base of an energy pyramid represents which organisms in an ecosystem?
    a. producers
    b. carnivores
    c. herbivores
    d. scavengers

12. Which of the following is the correct order in a food chain?
    a. sun → producers → herbivores → scavengers → carnivores
    b. sun → consumers → predators → parasites → hosts
    c. sun → producers → decomposers → consumers → omnivores
    d. sun → producers → herbivores → carnivores → scavengers

13. Remoras and sharks have a relationship best described as
    a. mutualism.
    b. commensalism.
    c. predator and prey.
    d. parasitism.

Short Answer

14. Briefly describe the habitat and niche of the gray wolf.

15. What might different species of trees in a forest compete for?

16. How do limiting factors affect the carrying capacity of an environment?

17. What is coevolution?

24

Concept Mapping

18. An answer to this exercise can be found at the front of this book.

CRITICAL THINKING AND PROBLEM SOLVING

19. An ecosystem with no decomposers could not exist very long because the ecosystem would soon become buried under the bodies of its dead organisms. Decomposers recycle dead material back into the ecosystem.

Concept Mapping

18. Use the following terms to create a concept map: individual organisms, producers, populations, ecosystems, consumers, herbivores, communities, carnivores, the biosphere.

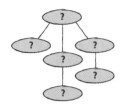

CRITICAL THINKING AND PROBLEM SOLVING

Write one or two sentences to answer the following questions:

19. Could a balanced ecosystem contain producers and consumers but no decomposers? Why or why not?

20. Some biologists think that certain species, such as alligators and wolves, help maintain biological diversity in their ecosystems. Predict what might happen to other species, such as gar fish or herons, if alligators were to become extinct in the Florida Everglades.

21. Does the Earth have a carrying capacity for humans? Explain your answer.

22. Explain why it is important to have a variety of organisms in a community of interacting species. Give an example.

MATH IN SCIENCE

23. The plants in each square meter of an ecosystem obtained 20,810 Calories of the sun's energy by photosynthesis per year. The herbivores in that ecosystem ate all of the plants, but they obtained only 3,370 Calories of energy. How much energy did the plants use for their own life processes?

INTERPRETING GRAPHICS

Examine the following graph, which shows the population growth of a species of *Paramecium*, a slipper-shaped, single-celled microorganism, over a period of 18 days. Food was occasionally added to the test tube in which the paramecia were grown. Answer the following questions:

24. What is the carrying capacity of the test tube as long as food is added?

25. Predict what will happen to the population if the researcher stops adding food to the test tube.

26. What keeps the number of *Paramecium* at a steady level?

27. Predict what might happen if the amount of water is doubled and the food supply stays the same.

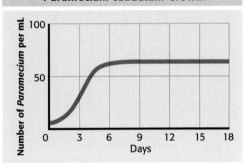

***Paramecium caudatum* Growth**

Reading Check-up

Take a minute to review your answers to the Pre-Reading Questions found at the bottom of page 2. Have your answers changed? If necessary, revise your answers based on what you have learned since you began this chapter.

20. A predator, such as the alligator, removes the weak, old, or sick individuals from its prey population and contributes to the health of that population. If alligators became extinct in the Everglades, gars and herons might soon overpopulate. They would weaken their own food sources and possibly contribute to the extinction of other species.

21. The Earth does have a carrying capacity for humans. We have extended our carrying capacity by growing more crops and controlling our environments to make previously uninhabitable areas livable. But we will eventually reach a carrying capacity that we cannot exceed.

22. It is important to have a variety of organisms in a community of interacting species because most species depend on other species for survival. All consumers depend on producers for food. For example, rabbits depend on grasses for food. Mice eat seeds and berries. A hawk depends on small animals like rabbits and mice for food and on trees for resting and nesting. A variety of producers supports a variety of consumers.

MATH IN SCIENCE

23. 17,440 Cal

INTERPRETING GRAPHICS

24. about 65
25. The population will crash due to starvation.
26. The limiting factors, such as food supply, limit growth.
27. Any single factor can be limiting. Space may be increased, but unless food is increased, the carrying capacity is the same.

Concept Mapping Transparency 18

Blackline masters of this Chapter Review can be found in the **Study Guide.**

Background

Follicle mites are harmless, but other kinds of mites are not benign guests. Mites are responsible for the devastation of honeybee populations in the United States. In the last few decades, tracheal mites and varroa mites have cut the number of domesticated colonies in half. There are almost no wild honeybees left in the United States.

The honeybee is a hardworking and important participant in agriculture. Apples and other orchard fruits require the honeybee for pollination. Domesticated honeybees annually enable the production of $10 billion worth of crops as well as produce $250 million worth of honey.

Scientists and beekeepers are trying various solutions to the mite infestations. Some chemicals seem to help. More promising solutions are found in new strains of honeybees bred to be resistant to mites, and in some honeybees that have developed the ability to groom mites off each other.

Health Watch

An Unusual Guest

What has a tiny tubelike body and short stumpy legs and lives upside down in your eyebrows and eyelashes? Would you believe a small animal? It's called a follicle mite, and humans are its host organism. Like all large animals, human beings are hosts to a variety of smaller creatures. They live in or on our bodies and share our bodies' resources. But none of our guests are stranger than follicle mites. They feed on oil and dead cells from your skin.

▲ A follicle mite is smaller than the period at the end of this sentence.

What Are They?

Follicle mites are arachnids—relatives of spiders. They are about 0.4 mm long, and they live in hair follicles all over your body. Usually they like to live in areas around the nose, cheek, forehead, chin, eyebrows, and eyelashes.

Follicle Mites Don't Bite

These tiny guests are almost always harmless, and they seldom live on children and adolescents. And you probably wouldn't even know they were there. Studies reveal that between 97 percent and 100 percent of all adults have these mites. Except in rare cases, follicle mites in adults are also pretty harmless.

Some Health Concerns

Although follicle mites rarely cause problems, they are sometimes responsible for an acnelike condition around the nose, eyebrows, and eyelashes. A large number of mites (up to 25) may live in the same follicle. This can cause an inflammation of the follicle. The follicle does not swell like acne; instead it becomes red and itchy.

Mites living in eyelashes and on eyelids can irritate those areas. The inflammation causes itchy eyelids or eyebrows. But such inflammations are rare, and the condition clears up very quickly when suitable medication is applied. So while follicle mites may be one of the strangest guests living on human skin, they are almost never a problem.

Other Companions

Many tiny organisms make their home in humans' bodies. Bacteria within the body may help maintain proper pH levels. Even *Escherichia coli*, a type of bacterium that can cause severe health problems, lives in the human colon. Without *E. coli*, a person would be unable to produce enough vitamin K or folic acid.

On Your Own

▶ Do some more research on follicle mites. Search for *Demodex folliculorum* or *Demodex brevis*. Find out more about some of the other strange organisms that rely on humans' bodies for food and shelter. Report on your findings, or write a story from the organism's point of view.

26

Answer to On Your Own

If you can find a copy of the book *Furtive Fauna: A Field Guide to the Creatures Who Live On You,* by Roger M. Knutson, you and your students will have more than enough information about follicle mites, tooth amoebas, bedbugs, ticks, and other organisms that find humans a wonderful place to live. There are a number of sites on the Internet that will provide additional information and links. (Use keywords such as follicle mites, Demodex, and human parasites.)

EYE ON THE ENVIRONMENT

Alien Invasion

A group of tiny aliens left their ship in Mobile, Alabama. Their bodies were red and shiny, and they walked on six legs. The aliens looked around and then quietly crawled off to make homes in the new land.

Westward Ho!

In 1918, fire ants were accidentally imported into the United States by a freighter ship from South America. In the United States, fire ants have no natural predators or competitors. In addition, these ants are extremely aggressive, and their colonies can harbor many queens, instead of just one queen, like many other ant species. With all these advantages, it is not surprising that the ants have spread like wildfire. By 1965, fire-ant mounds were popping up on the southeastern coast and as far west as Texas. Today they are found in at least 10 southern states and may soon reach as far west as California.

Jaws of Destruction

Imported fire ants have done a lot of damage as they have spread across the United States. Because they are attracted to electrical currents, they chew through wire insulation, causing shorts in electrical circuits. The invaders have also managed to disturb the natural balance of native ecosystems. In some areas, they have killed off 70 percent of the native ant species and 40 percent of other native insect species. Each year, about 25,000 people seek medical attention for painful fire-ant bites.

▲ Three types of fire ants are found in a colony: the queen, workers, and males. Notice how the queen ant dwarfs the worker ants.

Fighting Fire

Eighty years after the fire ants' introduction into the United States, the destructive ants continue to multiply. About 157 chemical products, including ammonia, gasoline, extracts from manure, and harsh pesticides, are registered for use against fire ants, but most have little or no success. Unfortunately, many of these remedies also harm the environment. By 1995, the government had approved only one fire-ant bait for large-scale use.

An Ant-Farm Census

▶ How many total offspring does a single fire-ant queen produce if she lives for 5 years and produces 1,000 eggs a day? If a mound contains 300,000 ants, how many mounds will her offspring fill?

EYE ON THE ENVIRONMENT
Alien Invasion

Background

In combating the spread of fire ants, people must be careful not to harm the environment. In the 1960s, ground-up corncobs and soybean oil were mixed with an ant poison and sprayed over huge areas of land. Studies have since shown that the poison used is toxic to many species. In addition, many researchers speculate that this strategy may have backfired because fire ants can recover from disasters much more quickly than other types of ants. After an area was treated with insecticides, the first ants that were able to reinhabit it were fire ants. Similar scenarios occurred in many different areas. As a result of these treatments, fire ants have become a serious ecological problem in many areas.

Discussion

Discuss with students some safe and natural ways that they can eliminate ants around their home. For instance, they can squeeze fresh lemon or lime juice into holes or cracks in their homes and then leave the peels where they have seen ants. They can also scatter aromatic substances, such as mint, red pepper, or paprika, around spots where ants enter homes. These substances repel ants.

Answers to An Ant-Farm Census

A typical queen may produce over 1.8 million eggs in her life:

1,000 eggs/day × 365 days/year × 5 years = 1,825,000 eggs.

Point out that a queen may not produce eggs every day.

You might also want to explain to the students that by dividing by 300,000 ants per mound, this number is equal to just over six mounds, or more than one new mound every year!

Chapter Organizer

CHAPTER ORGANIZATION	TIME MINUTES	OBJECTIVES	LABS, INVESTIGATIONS, AND DEMONSTRATIONS	
Chapter Opener pp. 28–29	45	National Standards: UCP 2, 4, SAI 1, 2, HNS 2, 3	**Start-Up Activity,** A Classroom Aquarium, p. 29	
Section 1 The Cycles of Matter	90	▶ Trace the cycle of water between the atmosphere, land, and oceans. ▶ Diagram the carbon cycle, and explain its importance to living things. ▶ Diagram the nitrogen cycle, and explain its importance to living things. UCP 1–4, SAI 1, 2, SPSP 4, 5; Labs UCP 1–3, SAI 1	**Demonstration,** The Water Cycle, p. 31 in ATE **QuickLab,** Combustion, p. 32 **Discovery Lab,** Nitrogen Needs, p. 130 **Datasheets for LabBook,** Nitrogen Needs	
Section 2 Ecological Succession	90	▶ Define *succession.* ▶ Contrast primary and secondary succession. UCP 1–4, SAI 1, SPSP 2; Labs UCP 1–4, SAI 1	**Making Models,** A Passel o' Pioneers, p. 38 **Datasheets for LabBook,** Passel o' Pioneers **Long-Term Projects & Research Ideas,** Smokey Says…	

See page **T23** *for a complete correlation of this book with the*

NATIONAL SCIENCE EDUCATION STANDARDS.

TECHNOLOGY RESOURCES

 Guided Reading Audio CD English or Spanish, Chapter 2

 One-Stop Planner CD-ROM with Test Generator

 CNN Science, Technology, & Society, Taking Earth's Pulse, Segment 21

CLASSROOM WORKSHEETS, TRANSPARENCIES, AND RESOURCES	SCIENCE INTEGRATION AND CONNECTIONS	REVIEW AND ASSESSMENT
Directed Reading Worksheet **Science Puzzlers, Twisters & Teasers**		
Transparency 67, The Water Cycle **Directed Reading Worksheet,** Section 1 **Transparency 141,** The Water Cycle **Transparency 68,** The Carbon Cycle **Critical Thinking Worksheet,** Pass the Salt, Please **Transparency 69,** The Nitrogen Cycle **Reinforcement Worksheet,** What Goes Around. . . .	**Environment Connection,** p. 31 **Connect to Earth Science,** p. 31 in ATE **MathBreak,** Gallons Galore, p. 33 **Apply,** p. 33 **Weird Science:** Weather from Fire, p. 44 **Eye on the Environment:** The Mysterious Dead Zone, p. 45	**Section Review,** p. 33 **Quiz,** p. 33 in ATE **Alternative Assessment,** p. 33 in ATE
Directed Reading Worksheet, Section 2 **Transparency 70,** Primary Succession **Transparency 71,** Secondary Succession	**Math and More,** p. 35 in ATE **Connect to Earth Science,** p. 35 in ATE **Multicultural Connection,** p. 35 in ATE	**Self-Check,** p. 36 **Homework,** p. 36 in ATE **Section Review,** p. 37 **Quiz,** p. 37 in ATE **Alternative Assessment,** p. 37 in ATE

internet connect

  Holt, Rinehart and Winston On-line Resources

go.hrw.com

For worksheets and other teaching aids related to this chapter, visit the HRW Web site and type in the keyword: **HSTCYC**

 National Science Teachers Association

www.scilinks.org

Encourage students to use the *sci*LINKS numbers listed in the internet connect boxes to access information and resources on the **NSTA** Web site.

END-OF-CHAPTER REVIEW AND ASSESSMENT

Chapter Review in Study Guide
Vocabulary and Notes in Study Guide
Chapter Tests with Performance-Based Assessment, Chapter 2 Test
Chapter Tests with Performance-Based Assessment, Performance-Based Assessment 2
Concept Mapping Transparency 19

Chapter Resources & Worksheets

Visual Resources

TEACHING TRANSPARENCIES

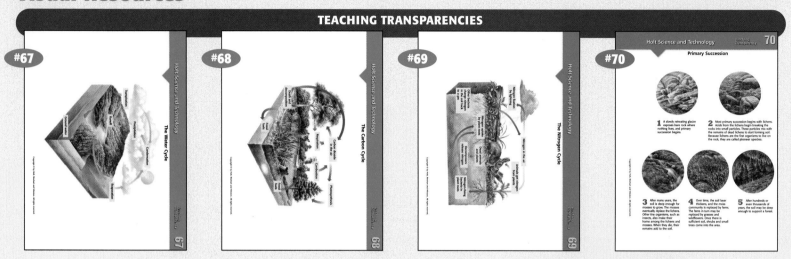

#67 — Holt Science and Technology — The Water Cycle

#68 — Holt Science and Technology — The Carbon Cycle

#69 — Holt Science and Technology — The Nitrogen Cycle

#70 — Holt Science and Technology — Teaching Transparency 70 — Primary Succession

1 A slowly retreating glacier exposes bare rock where nothing lives, and primary succession begins.

2 Most primary succession begins with lichens. Acids from the lichens begin breaking the rocks into small particles. These particles mix with the remains of dead lichens to start forming soil. Because lichens are the first organisms to live on the rock, they are called pioneer species.

3 After many years, the soil is deep enough for mosses to grow. The mosses eventually replace the lichens. Other tiny organisms, such as insects, also make their home among the lichens and mosses. When they die, their remains add to the soil.

4 Over time, the soil layer thickens, and the moss community is replaced by ferns. The ferns in turn may be replaced by grasses and wildflowers. Once there is sufficient soil, shrubs and small trees come into the area.

5 After hundreds or even thousands of years, the soil may be deep enough to support a forest.

TEACHING TRANSPARENCIES

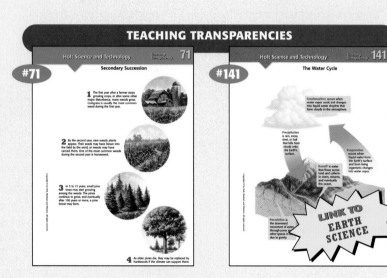

#71 — Holt Science and Technology — Teaching Transparency 71 — Secondary Succession

1 The first year after a farmer stops growing crops, or after some other major disturbance, many weeds grow. Crabgrass is usually the most common weed during the first year.

2 By the second year, new weedy plants appear. Their seeds may have blown into the field by the wind, or insects may have carried them. One of the most common weeds during the second year is horseweed.

3 In 5 to 15 years, small pine trees may start growing among the weeds. The pines continue to grow, and eventually after 100 years or more, a pine forest may form.

4 As older pines die, they may be replaced by hardwoods if the climate can support them.

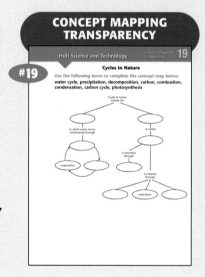

#141 — Holt Science and Technology — Teaching Transparency 141 — The Water Cycle

Condensation occurs when water vapor cools and changes into liquid water droplets that form clouds in the atmosphere.

Precipitation is rain, snow, sleet, or hail that falls from clouds onto the Earth's surface.

Evaporation occurs when liquid water from the Earth's surface and from living organisms changes into water vapor.

Runoff is water that flows across land and collects in rivers, streams, and eventually the ocean.

Percolation is the downward movement of water through pores and other spaces in soil due to gravity.

LINK TO EARTH SCIENCE

CONCEPT MAPPING TRANSPARENCY

#19 — Holt Science and Technology — Concept Mapping Transparency 19 — Cycles in Nature

Use the following terms to complete the concept map below: **water cycle, precipitation, decomposition, carbon, combustion, condensation, carbon cycle, photosynthesis**

Meeting Individual Needs

DIRECTED READING

#2 — DIRECTED READING WORKSHEET — *Cycles in Nature*

Chapter Introduction
As you begin this chapter, answer the following.
1. Read the title of the chapter. List three things that you already know about this subject.

2. Write two questions about this subject that you would like answered by the time you finish this chapter.

Section 1: The Cycles of Matter (p. 30)
3. The matter in your body has been on Earth only since the time you were born. True or False? (Circle one.)
4. All of the following are true about matter EXCEPT
   a. it occupies space and has mass.
   b. it is used over and over again.
   c. only living things contain matter.
   d. each kind of matter has its own cycle.

The Water Cycle (p. 30)
5. The water cycle is the movement of water among the atmosphere, the land, the ocean, and _____

REINFORCEMENT & VOCABULARY REVIEW

#2 — REINFORCEMENT WORKSHEET — *What Goes Around . . .*

Complete this worksheet after reading Chapter 19, Section 1.
Diagrams of the carbon cycle and water cycle are shown below and on the next page. The opposing processes in both cycles are represented as arrows. Answer the questions relating to each diagram.

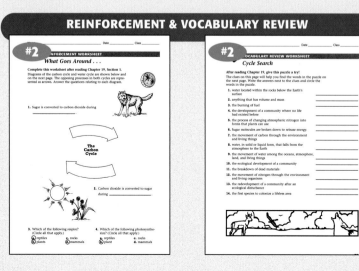

1. Sugar is converted to carbon dioxide during _____

The Carbon Cycle

2. Carbon dioxide is converted to sugar during _____

3. Which of the following respire? (Circle all that apply.)
   a. reptiles
   b. plants
   c. rocks
   d. mammals

4. Which of the following photosynthesize? (Circle all that apply.)
   a. reptiles
   b. plant
   c. rocks
   d. mammals

#2 — VOCABULARY REVIEW WORKSHEET — *Cycle Search*

After reading Chapter 19, give this puzzle a try!
The clues on this page will help you find the words in the puzzle on the next page. Write the answers next to the clues and circle the words in the puzzle.

1. water located within the rocks below the Earth's surface
2. anything that has volume and mass
3. the burning of fuel
4. the development of a community where no life had existed before
5. the process of changing atmospheric nitrogen into forms that plants can use
6. Sugar molecules are broken down to release energy.
7. the movement of carbon through the environment and living things
8. water, in solid or liquid form, that falls from the atmosphere to the Earth
9. the movement of water among the oceans, atmosphere, land, and living things
10. the ecological development of a community
11. the breakdown of dead materials
12. the movement of nitrogen through the environment and living organisms
13. the redevelopment of a community after an ecological disturbance
14. the first species to colonize a lifeless area

SCIENCE PUZZLERS, TWISTERS & TEASERS

#2 — SCIENCE PUZZLERS, TWISTERS & TEASERS — *Cycles in Nature*

Nature is A"maze"ing
1. Six words associated with cycles in nature are hidden in the maze below. Each word contains at least 10 letters. Locate the words by tracing one of the six different paths that lead into and out of the maze. The letters of all of the words occur in order along each path, none of the paths cross one another, and all of the letters are used once and only once to spell the six words. List the words on the lines provided.

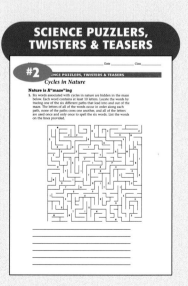

Cycles in Nature

Chapter 2 • Cycles in Nature

Review & Assessment

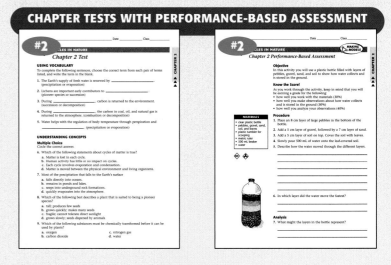

STUDY GUIDE

#2 VOCABULARY & NOTES WORKSHEET

Cycles in Nature

By studying the Vocabulary and Notes listed for each section below, you can gain a better understanding of this chapter.

SECTION 1

Vocabulary

In your own words, write a definition for the following terms in the space provided.

1. precipitation

2. evaporation

3. ground water

4. decomposition

5. combustion

Notes

Read the following section highlights. Then, in your own words, write the highlights in your ScienceLog.

- Materials used by living things continually cycle through ecosystems.
- In the water cycle, water moves through the ocean, atmosphere, land, and living things.
- Precipitation, transpiration, evaporation, and condensation are important processes in the water cycle.
- Water that falls is held in soil or porous rocks as ground water.
- Photosynthesis, respiration, decomposition, and combustion are important steps in the carbon cycle.
- Carbon enters plants from the nonliving environment as carbon dioxide.
- The process of changing nitrogen gas into forms that plants can use is called nitrogen fixation.

#2 CHAPTER REVIEW WORKSHEET

Cycles in Nature

USING VOCABULARY

To complete the following sentences, choose the correct term from each pair of terms listed below, and write the term in the space provided.

1. During _____ water moves from the atmosphere to the land and ocean. (evaporation or precipitation)

2. All biological molecules contain _____. (carbon or carbon dioxide)

3. The combustion of coal, oil, and natural gas is part of the _____. (nitrogen cycle or carbon cycle)

4. The development of a community on bare, exposed rock is an example of _____. (primary succession or secondary succession)

5. The recovery of Yellowstone National Park following the fires of 1988 is an example of _____. (primary succession or secondary succession)

UNDERSTANDING CONCEPTS

Multiple Choice

6. Water changes from a liquid to a vapor during
   a. precipitation.   c. evaporation.
   b. respiration.   d. decomposition.

7. The process of burning fuel, such as oil and coal, is
   a. combustion.   c. decomposition.
   b. respiration.   d. photosynthesis.

8. One of the most common plants in a recently abandoned farm field is
   a. horseweed.   c. young oak and hickory trees.
   b. young pine trees.   d. crabgrass.

9. Which of the following statements about ground water is true?
   a. It stays underground for a few days.
   b. It is stored in underground caverns or porous rock.
   c. It is salty like ocean water.
   d. It never reenters the water cycle.

10. Which of the following processes produces carbon dioxide?
    a. decomposition   c. combustion
    b. respiration   d. All of the above

11. During nitrogen fixation, nitrogen gas is converted into a form that _____ can use.
    a. plants   c. fungi
    b. animals   d. All of the above

CHAPTER TESTS WITH PERFORMANCE-BASED ASSESSMENT

#2 CYCLES IN NATURE

Chapter 2 Test

USING VOCABULARY

To complete the following sentences, choose the correct term from each pair of terms listed, and write the term in the blank.

1. The Earth's supply of fresh water is renewed by _____. (precipitation or evaporation)

2. Lichens are important early contributors to _____. (pioneer species or succession)

3. During _____ carbon is returned to the environment. (succession or decomposition)

4. During _____ the carbon in coal, oil, and natural gas is returned to the atmosphere. (combustion or decomposition)

5. Water helps with the regulation of body temperature through perspiration and _____. (precipitation or evaporation)

UNDERSTANDING CONCEPTS

Multiple Choice

Circle the correct answer.

6. Which of the following statements about cycles of matter is true?
   a. Matter is lost in each cycle.
   b. Human activity has little or no impact on cycles.
   c. Each cycle involves evaporation and condensation.
   d. Matter is moved between the physical environment and living organisms.

7. Most of the precipitation that falls to the Earth's surface
   a. falls directly into oceans.
   b. remains in ponds and lakes.
   c. seeps into underground rock formations.
   d. quickly evaporates into the atmosphere.

8. Which of the following best describes a plant that is suited to being a pioneer species?
   a. tall; produces few seeds
   b. grows quickly; makes many seeds
   c. fragile; cannot tolerate direct sunlight
   d. grows slowly; seeds dispersed by animals

9. Which of the following substances must be chemically transformed before it can be used by plants?
   a. oxygen   c. nitrogen gas
   b. carbon dioxide   d. water

#2 CYCLES IN NATURE **MAKING MODELS**

Chapter 2 Performance-Based Assessment

Objective

In this activity you will use a plastic bottle filled with layers of pebbles, gravel, sand, and soil to show how water collects and is stored in the ground.

Know the Score!

As you work through the activity, keep in mind that you will be earning a grade for the following:

- how well you work with the materials (30%)
- how well you make observations about how water collects and is stored in the ground (30%)
- how well you analyze your observations (40%)

MATERIALS

- clear plastic bottle
- pebbles, gravel, sand, soil, and leaves
- plastic tumbler for scooping
- metric ruler
- 500 mL beaker
- water

Procedure

1. Place an 8 cm layer of large pebbles in the bottom of the bottle.
2. Add a 5 cm layer of gravel, followed by a 7 cm layer of sand.
3. Add a 5 cm layer of soil on top. Cover the soil with leaves.
4. Slowly pour 500 mL of water onto the leaf-covered soil.
5. Describe how the water moved through the different layers.

6. In which layer did the water move the fastest?

Analysis

7. What might the layers in the bottle represent?

Lab Worksheets

LONG-TERM PROJECTS & RESEARCH IDEAS

#2 STUDENT WORKSHEET **DISCOVERY LAB**

Smokey Says . . .

"Remember, only you can prevent forest fires."

For over 50 years, Smokey the Bear's words have reminded Americans to protect forests from fires. You might think that Smokey was created by environmentalists. Actually, he was invented to help the American war effort during World War II. The government needed trees to build ships and airplanes, but trees in the forests were going up in smoke, mostly because of careless people. So Smokey was "born" to make people more careful about fires when visiting forests.

Up in Smoke

1. Forest fires are bad, right? Well, not necessarily. In the early 1990s, drought caused forest fires to rage throughout areas of the Pacific Northwest. Modern firefighting techniques helped to contain these fires. But some ecologists think that forest fires should be allowed to burn unless humans and their homes are threatened. They claim that fires are part of the natural process of succession and renewal. Research both sides of this debate. Which argument do you agree with? Take one position and write a newspaper editorial defending your position.

Other Research Ideas

2. In 1992, Hurricane Andrew blew through southern Dade County, in Florida, and caused a lot of destruction. Investigate the stage of succession that this area has now reached. What plants and animals have returned to the area? Investigate another area that has experienced devastation due to a natural disaster. Compare the current state of succession of this region with the area that was damaged by Hurricane Andrew. Use your research to write a fictional interview with a rock, tree, or river that survived one of the disasters and is watching succession happen.

Long-Term Project Idea

3. If you watch carefully, you might just see succession in action. Find a site that shows signs of primary and secondary succession. Photograph the changes over the course of several months and then create a display exhibiting and discussing the changes.

USEFUL TERMS
primary succession
secondary succession

DATASHEETS FOR LABBOOK

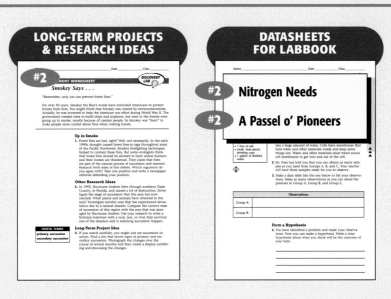

#2 Nitrogen Needs

#2 A Passel o' Pioneers

- 1 box of salt
- small, clear-plastic drinking cups
- 1 gallon of distilled water

tain a large amount of water. Cells have membranes that hold water and other materials inside and keep some things out. Water and other materials must travel across cell membranes to get into and out of the cell.

2. Mr. Fries has told you that you can obtain as many samples as you need from Groups A, B, and C. Your teacher will have these samples ready for you to observe.

3. Make a data table like the one below to list your observations. Make as many observations as you can about the potatoes in Group A, Group B, and Group C.

Observations	
Group A:	
Group B:	

Form a Hypothesis

4. You have identified a problem and made your observations. Now you can make a hypothesis. Write a clear hypothesis about what you think will be the outcome of your tests.

Applications & Extensions

CRITICAL THINKING & PROBLEM SOLVING

#2 CRITICAL THINKING WORKSHEET

Pass the Salt, Please

The following excerpts describe an experimental means of food production:

Earth may be the Ocean Planet but most terrestrial creatures—including humans—depend for food on plants irrigated by freshwater from rainfall, rivers, lakes, springs and streams. None of the top five plants eaten by people—wheat, corn, rice, potatoes and soybeans—can tolerate salt; expose them to seawater, and they droop, shrivel and die within days. . . . We have tested the feasibility of seawater agriculture and have found that it works well in the sandy soils of desert environments. Seawater agriculture is defined as growing salt-tolerant crops on land using water pumped from the ocean for irrigation. . . .

Seawater agriculture must fulfill two requirements to be cost-effective. First, it must produce useful crops at yields high enough to justify the expense of pumping irrigation water from the sea. Second, researchers must develop agronomic techniques for growing seawater-irrigated crops in a sustainable manner—one that doesn't damage the environment.

From "Irrigating Crops with Seawater" by Edward P. Glenn, J. Jed Brown and James W. O'Leary from *Scientific American*, August 1998. Available online at http://www.sciam.com/1998/0898issue/0898glenn.html. Reprinted by permission of *Edward P. Glenn, J. Jed Brown, and James W. O'Leary*.

USEFUL TERMS
feasibility
possible use of
yields
amounts
agronomic
having to do with crop and farmland management
agronomy
crop and farmland management

Making Inferences

1. Why are scientists interested in irrigating crops with sea water?

Demonstrating Reasoned Judgment

2. Pumping sea water to an inland destination is costly. Why?

SCIENCE TECHNOLOGY

#21 Science in the News: Critical Thinking Worksheets

Segment 21

Taking Earth's Pulse

1. Why is it important to monitor changes in plant life in the ocean?

2. How do you think organisms in the ocean would respond to lower levels of phytoplankton in certain areas?

3. How do you think the satellite discussed in the video could be used to protect the environment?

4. What purpose could the satellite serve in the case of an oil spill?

The Cycles of Matter

▶ The Water Cycle

Billions of years ago, Earth was a mass of dust and hot gases. Among these gases were the elements hydrogen and oxygen. The two elements combined to form water. In the early water cycle, the temperature on Earth was too hot for water to exist as a liquid or a solid. Dense clouds of water vapor surrounded Earth, and droplets that fell to Earth evaporated immediately.

- After millions of years, the temperature on Earth cooled enough for water to exist as a liquid. This cooling triggered massive rainfall. Huge amounts of water vapor in the atmosphere condensed, creating torrential downpours. This first rainfall—scientists believe it lasted for hundreds of years—filled depressions in Earth's surface, forming the first lakes, rivers, and oceans.

- Not all water molecules are alike. There are six different isotopes of oxygen, but because of their relative abundance, drinking water primarily contains one. There are three isotopes of hydrogen. Known as protium (single-weight hydrogen), deuterium (double-weight hydrogen), and tritium (triple-weight hydrogen), these isotopes can be combined in many different ways to create water molecules of different weights. When combined with oxygen, protium forms light water, deuterium forms heavy water, and tritium forms superheavy water. In nature, most water is light.

IS THAT A FACT!

- ☛ The transpiration of water by plants contributes a huge amount of water vapor to the air. A fully grown oak tree, for example, may transpire up to 380 L of water per day!

▶ The Carbon Cycle

Like water, carbon is recycled in the natural world. Respiration among living things releases carbon dioxide, which is taken up by plants and converted to sugars. The carbon that accumulates in plants as carbohydrates circulates in nature in a number of ways. When plants are eaten, their carbon is transferred to animals. The waste products of the animals' metabolism—including carbon dioxide—are returned to nature. Finally, when living things die, their decomposition releases carbon.

- Human activities can alter the balance of the global carbon cycle. The burning of fossil fuels, for example, releases vast amounts of carbon dioxide into the atmosphere. The carbon contained in fossil fuels is formed very slowly, yet its combustion is rapid. As a result, the amount of carbon dioxide in the atmosphere is rising. This is of concern to some scientists because carbon dioxide is a greenhouse gas. Carbon dioxide, along with other greenhouse gases, traps heat and prevents it from radiating from Earth. Scientists are concerned that this phenomenon is causing the temperature on Earth to rise.

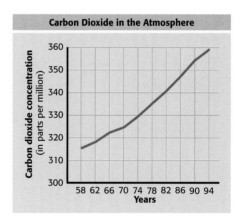

Carbon Dioxide in the Atmosphere

Ecological Succession

▶ The Nature of Ecosystems

Like living things, ecosystems are born, develop, and mature. Events such as fires, storms, and volcanic eruptions can disrupt ecosystems, setting into action a chain of events known as succession. Though discussions of succession generally center on forests, all ecosystems undergo the process.

- Over the last several thousand years, the water level of the Great Lakes has gradually decreased. As the lake shores expanded, sand dunes formed. These dunes are ecosystems. Like a forest ecosystem, a sand dune ecosystem follows a pattern of succession.  Sand is unstable, infertile, and inhospitable. At first, only the hardiest plants can survive there. These pioneer plants stabilize the sand, enriching the soil as they decay. This makes a much more hospitable environment for other living things. Grasses move into the area, covering the dune. Their extensive root systems further stabilize the sand, holding the dune in place. Shrubs and pines then join the ecosystem, and the shade they produce eventually kills the grasses. A thick pine forest eventually replaces the grasses because the shade produced by the pines makes them unable to thrive. The pines create a more hospitable environment for deciduous trees. The pines are gradually replaced by trees such as oaks, maples, and elms. These trees are able to reproduce in the shady forest environment.

▶ Forest Management

The branch of forestry concerned with the life cycle of trees and the forest is known as forest management. Forest managers manage the forest for economic benefits and protect the ecosystem for wildlife. Balancing the two responsibilities is sometimes difficult.

- Forest managers perform some of the same functions carried out by nature. They plant trees, thin forest stands, and remove trees. Unmanaged trees burn or rot, so forest managers usually find ways to use the wood of managed forests.

- One controversial activity carried out by forest managers is known as prescribed burning. Though dangerous and often destructive, fire is a necessary component of the life cycle of forest communities. Its effects on wildlife are complex; fire can assist in the recycling of nutrients, regulate plant succession, help maintain biological diversity, reduce biomass, and help control insect and disease populations. In 1998, the U.S. Forest Service burned between 850,000 and 1 million acres.

▶ Forest Wildlife

As forests undergo succession, they attract different forms of wildlife. Piney forests attract different species than hardwood forests do. This is because the amount and type of resources in each favors certain animals but discourages others. For example, the southern pine beetle can survive only in pine forests. Other animals, known as browsers, can live in either hardwood or pine forest habitat. These animals, including deer and elk, rely upon the forest for shelter but often find food on its periphery.

▶ Plant Competition and Succession

Plant succession is affected by competition among plants for resources. Plants compete with one another for sunlight, water, and nitrogen, among other resources. As with other living things, plants that are able to efficiently use these resources are able to grow and reproduce more effectively than their competitors.

For background information about teaching strategies and issues, refer to the *Professional Reference for Teachers.*

Cycles in Nature

 Pre-Reading Questions

Students may not know the answers to these questions before reading the chapter, so accept any reasonable response.

Suggested Answers

1. Recycling means the process of making another product from a reprocessed product. Water, nitrogen, and carbon are three of the many substances essential to life on Earth that cycle from the physical environment, through living things, and back to the physical environment in different forms.

2. Rain falls to the Earth as precipitation. If it falls into the ocean, it may evaporate and fall as precipitation again. If rainwater falls on land, it may join a stream, river, or other body of water. It may also trickle down into the Earth to become ground water. It will eventually reenter the atmosphere through evaporation or transpiration.

CHAPTER
2

Cycles in Nature

Sections

Pre-Reading
Questions

1. What is meant by *recycling*?

2. What happens to rainwater after it falls to Earth?

DESERT POST

Ever tried to send a desert to anyone?
Thanks to the U.S. Postal Service, you
can! Stamps were made from this picture of the Sonoran Desert. The stamps
are intended to promote a greater appre-
ciation of the diversity of the Sonoran
Desert. The desert scene was the first of
a series that commemorates America's
natural environment. In this chapter,
you will learn about the natural environment and how it works.

A CLASSROOM AQUARIUM

Did you know an aquarium is a
small environment? In this activity,
you will put an aquarium together.
As you plan the aquarium, think
about how all the parts are connected with each other.

Procedure

1. Get a **tank** from your teacher.
Check the Internet, your library,
or a pet store to find directions
on the proper way to clean and
prepare an aquarium.

2. Find out about the kinds of **plants**
and **animals** that you can put in
your aquarium.

3. Choose a place to put the tank,
and tell your teacher your plans.
Then set up the aquarium.

Analysis

4. How is the aquarium similar to
and different from a natural body
of water? Identify the limitations of
your model.

5. After you read Section 1, take
another look at the aquarium. See
if you can name all the parts of
this small ecosystem.

29

START-UP Activity

A CLASSROOM AQUARIUM

MATERIALS
FOR EACH GROUP: • aquarium • various aquatic plants • aquatic animals

Answers to START-UP Activity

4. Sample answer: The aquarium
is similar to a natural body of
water because it is an aquatic
environment that contains living
things. It is different from a
natural body of water in that it
is confined by glass walls and
may receive a different amount
of sunlight. Some limitations of
this model include the very
limited number of organisms in
the aquarium, the glass walls,
the different nutrient inputs and
outputs, and the lack of contact
with other environments.

5. Sample answer: the water, the
plants, the animals, the tank,
the gravel (or stones), the
filtration system, the air pump

Focus

The Cycles of Matter

This section explains that matter is neither created nor destroyed. The water, carbon, and other materials in the environment are constantly recycled. The water cycle and the carbon cycle are used to illustrate how matter is recycled in nature.

 **Bellringer**

Ask students to answer the following question in their ScienceLog:

> What are the four major steps of the water cycle?
> (transpiration, evaporation, condensation, and precipitation)

1 Motivate

DISCUSSION

Initiate a discussion with students about the properties of matter. Remind them that matter is anything that has mass and takes up space. Make sure that students understand that matter is always conserved. Illustrate this concept by telling them that the water they used to brush their teeth in the morning may have once been a dinosaur's bathwater!

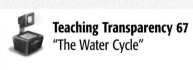
Teaching Transparency 67
"The Water Cycle"

Directed Reading Worksheet Section 1

 internet connect

SCi**LINKS**
NSTA

TOPIC: The Water Cycle
GO TO: www.scilinks.org
*sci*LINKS **NUMBER:** HSTL455

Terms to Learn

precipitation   decomposition
evaporation   combustion
ground water

What You'll Do

◆ Trace the cycle of water between the atmosphere, land, and oceans.
◆ Diagram the carbon cycle, and explain its importance to living things.
◆ Diagram the nitrogen cycle, and explain its importance to living things.

The Cycles of Matter

The matter in your body has been on Earth since the planet was formed billions of years ago! *Matter,* which is anything that occupies space and has mass, is used over and over again. Each kind of matter has its own cycle. In these cycles, matter moves among the environment and living things.

The Water Cycle

The movement of water among the oceans, atmosphere, land, and living things is known as the *water cycle.* Locate each part of the water cycle in **Figure 1** as it is discussed.

Precipitation Water moves from the atmosphere to the land and oceans as **precipitation,** which includes rain, snow, sleet, and hail. About 91 percent of precipitation falls into the ocean. The rest falls on land, renewing the supply of fresh water.

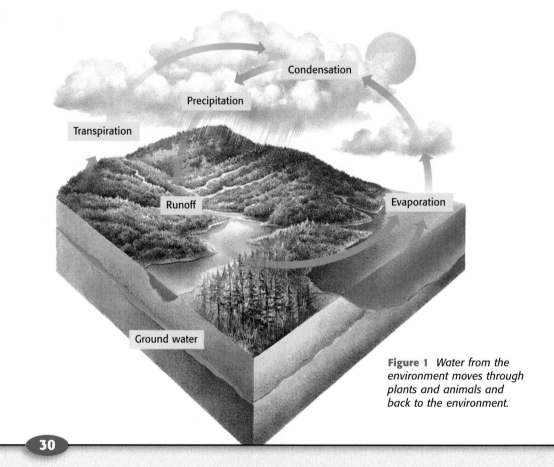

Condensation

Precipitation

Transpiration

Runoff

Evaporation

Ground water

Figure 1 *Water from the environment moves through plants and animals and back to the environment.*

WEIRD SCIENCE

Converting ice to liquid requires thermal energy. When the process is reversed, some thermal energy is given off. Farmers use this principle when they spray crops with water during sudden cold weather. As the water freezes on the fruit, heat is given off, which prevents the fruit from freezing.

SCIENCE HUMOR

Q: If H_2O is water, what is H_2O_4?

A: Drinking!

Evaporation Water cycles back to the atmosphere through evaporation. During **evaporation,** the sun's heat causes water to change from liquid to vapor. When the water vapor cools during the process of *condensation,* it forms a liquid that can fall to the Earth as precipitation.

Ground Water Some precipitation seeps into the ground, where it is stored in underground caverns or in porous rock. This water, known as **ground water,** may stay in the ground for hundreds or even thousands of years. Ground water provides water to the soil, streams, rivers, and oceans.

Water and Life All organisms, from tiny bacteria to animals and plants, contain a lot of water. Your body is composed of about 70 percent water. Water carries waste products away from body tissues. Water also helps regulate body temperature through perspiration and evaporation, returning water to the environment in a process called *transpiration.* Without water, there would be no life on Earth.

The Carbon Cycle

Carbon is essential to living things because it is part of all biological molecules. The movement of carbon from the environment into living things and back into the environment is known as the *carbon cycle,* shown in **Figure 2.**

Environment
C O N N E C T I O N

Carbon dioxide is being released into the atmosphere in increasing quantities. Carbon dioxide causes the atmosphere to hold heat. The warmer atmosphere causes the temperatures of the land and ocean to rise. This is known as global warming.

Figure 2 *Carbon may remain in the environment for millions of years before becoming available to living things.*

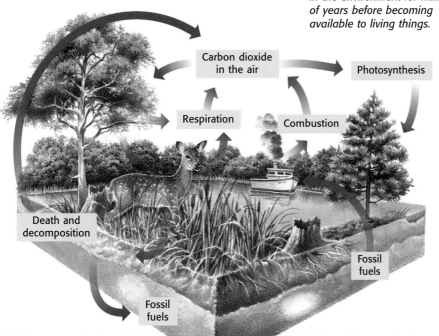

31

IS THAT A FACT!

A lush oasis can exist in a desert that seldom receives rain. Water beneath the desert, called fossil water, slowly seeps to the surface at the oasis, allowing plants and animals to live in an otherwise dry place. Fossil water is water that was trapped underground, away from the water cycle, tens of thousands of years ago. When the fossil water is gone, the oasis it feeds will disappear.

MATERIALS

- candle
- glass jar with lid
- modeling clay

Safety Caution: Remind students to review all safety cautions and icons before beginning this lab activity. Students should stand several meters back from the candle flame.

Answers to QuickLab

- The substance on the side of the jar is water. The warm air contains moisture, which condenses when it contacts the cool glass.
- The substance inside the jar is soot. Soot is carbon and is the product of the combustion of the candle wax and the wick.

Nitrogen Needs

RESEARCH

Writing Encourage interested students to use library or Internet resources to investigate artificial nitrogen fixation. Have them write brief reports of the methods of fixation and their applications, and allow time for them to report their findings to the class.

Teaching Transparency 69
"The Nitrogen Cycle"

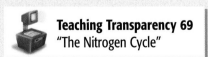
internetconnect

SCiLINKS **TOPIC:** The Nitrogen Cycle
GO TO: www.scilinks.org
sciLINKS NUMBER: HSTL465

QuickLab

Combustion

Place a **candle** on a **jar lid** and secure it with **modeling clay.** Light the candle. Hold the **jar** very close to the candle flame. What is deposited on the jar? Where did the substance come from? Now place the jar over the candle. What is deposited inside the jar? Where did this substance come from?

Photosynthesis Photosynthesis is the process by which carbon cycles from the environment into living things. During photosynthesis, plants use carbon dioxide from the air to make sugars. Most animals get the carbon they need by eating plants.

Respiration How does carbon return to the environment? Animals and plants both respire. During *respiration,* sugar molecules are broken down to release energy. Carbon dioxide and water are released as byproducts.

Decomposition The breakdown of dead materials into carbon dioxide and water is called **decomposition.** When fungi and bacteria decompose organic matter, they return carbon to the environment.

Combustion The carbon in coal, oil, and natural gas returns to the atmosphere as carbon dioxide when these fuels are burned. The process of burning fuel is known as **combustion.** Combustion provides much of the fuel people need to drive cars, heat homes, and make electricity.

The Nitrogen Cycle

The movement of nitrogen from the environment to living things and back again is called the *nitrogen cycle,* shown in **Figure 3.**

Figure 3 *Without bacteria, nitrogen could not enter living things or be returned to the atmosphere.*

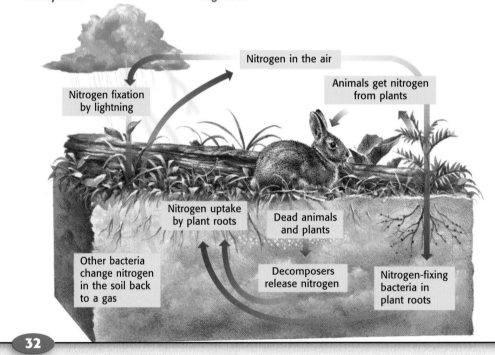

Nitrogen in the air

Nitrogen fixation by lightning

Animals get nitrogen from plants

Nitrogen uptake by plant roots

Dead animals and plants

Other bacteria change nitrogen in the soil back to a gas

Decomposers release nitrogen

Nitrogen-fixing bacteria in plant roots

32

MISCONCEPTION ///ALERT\\\

Students may believe that climate changes are caused solely by human activities that release carbon into the atmosphere. While it is important to be aware of our impact on the atmosphere, tell students most climate changes are not solely the result of human action. Scientists theorize that climate changes involve the complex interactions of natural phenomena, such as volcanic eruptions; changes in the output of the sun's energy; and continental drift. Deforestation and the burning of fossil fuels are just two possible factors affecting climates.

A Sea of Nitrogen About 78 percent of the Earth's atmosphere is nitrogen gas. However, most organisms cannot use nitrogen gas to obtain the nitrogen they need to build proteins and DNA. But bacteria in the soil are able to change nitrogen gas into forms that can be used by plants. This is called *nitrogen fixation.* Most animals get the nitrogen they need by eating plants.

Back to Gas The final step of the nitrogen cycle is also performed by bacteria in the soil. These bacteria are different species than the bacteria that fix nitrogen. The bacteria break down dead organisms and animal wastes. This process produces nitrogen gas, which is returned to the atmosphere.

MATH BREAK

Gallons Galore

An average person in the United States uses about 78 gal of water each day. How many liters is this? How many cubic centimeters? Remember: 1 gal = 3.79 L and 1 mL = 1 cm³.

The Pollution Cycle

Isabel read an article about how power plants near her home emit sulfur dioxide into the atmosphere. When sulfur dioxide mixes with water, it forms sulfuric acid, which is extremely toxic to living things. Isabel also learned that sulfur dioxide from these power plants has killed all the fish in a lake hundreds of kilometers away. Trees growing near the lake were also killed. Using what you know about the water cycle, write a letter to Isabel explaining how this could happen.

SECTION REVIEW

1. How are precipitation, evaporation, and ground water involved in the water cycle?

2. Draw a simple diagram of the nitrogen cycle. Make sure you include how animals get nitrogen.

3. **Analyzing Relationships** How is decomposition related to the carbon cycle?

internet connect

SCiLINKS
NSTA

TOPIC: The Water Cycle,
The Nitrogen Cycle
GO TO: www.scilinks.org
***sci*LINKS NUMBER:** HSTL455, HSTL465

33

▼ **Answers to Section Review**

1. In the water cycle, water falls to Earth's surface as precipitation. Some water reenters the atmosphere by evaporation and by transpiration from plants. Water condenses and falls again as precipitation.

2. Student diagrams should resemble **Figure 3.**

3. Living organisms consume and store carbon. With the help of bacteria, dead organisms decompose and release that stored carbon back into the environment.

Quiz

1. Compare the water cycle with the carbon cycle. What do both have in common? (Answers will vary but should reflect an understanding that, in each cycle, matter is used over and over again.)

2. Describe the role of living things in the carbon cycle. (Answers will vary. Students should note that plants obtain carbon from the atmosphere, using it to make sugars during photosynthesis. When animals eat the plants, they take in the carbon. Both plants and animals release carbon dioxide during cellular respiration.)

ALTERNATIVE ASSESSMENT

Writing **Concept Mapping** Have students create a concept map using the new terms from this section, the section title, and any necessary additional terms. Remind them to include apt connecting words and phrases to link the terms of the map.

Answer to APPLY

The letter to Isabel should state that sulfur dioxide could be carried by wind to the area near the lake. Water vapor from evaporation, transpiration, and rain mix with sulfur dioxide and can form sulfuric acid. When the acidic water falls as rain or other precipitation, it can contaminate the lake and harm or kill trees.

Focus

Ecological Succession

In this section, students are introduced to the development of communities by succession. Students will learn the difference between primary and secondary succession and the organisms characteristic of each. The gradual formation of a stable community is examined.

Bellringer

Ask students to imagine that they have been hired to oversee the maintenance of a forest. Have them answer the following questions to describe how they would approach the task:

- How would you evaluate the health of the forest?
- What actions would you take to keep the forest healthy?
- What factors might pose a threat to the health of the forest? How would you prevent them from causing harm?

1) Motivate

DISCUSSION

Tell students that in deciduous forest areas, hardwood trees sometimes replace the pines that come before them. Point out that pine forests provide valuable resources. Encourage students to speculate about how forest managers might prevent hardwood trees from replacing pine trees in managed forests.

Terms to Learn

succession
pioneer species

What You'll Do

◆ Define *succession*.
◆ Contrast primary and secondary succession.

Ecological Succession

Imagine you have a time machine that can take you back to the summer of 1988. If you had visited Yellowstone National Park during that year, you would have found large areas of the park burned to the ground. When the fires were put out, a layer of gray ash blanketed the forest floor. Most of the trees were dead, although many of them were still standing, as shown in **Figure 4.**

Figure 4 *Parts of Yellowstone National Park burned in 1988.*

Figure 5 *In the spring of 1989, regrowth was evident in the burned parts of Yellowstone National Park.*

Regrowth of a Forest

The following spring, the appearance of the "dead" forest began to change. In **Figure 5,** you can see that some of the dead trees are beginning to fall over, and small, green plants have begun to grow in large numbers. National Park foresters report that the number and kinds of plants growing in the recovering area have increased each year since the fire.

A gradual development of a community over time, such as the regrowth of the burned areas of Yellowstone National Park, is called **succession.** Succession takes place in all communities, not just those affected by disturbances such as forest fires. Succession occurs through predictable stages over time, as described on the following pages.

34

Directed Reading Worksheet Section 2

internetconnect

SCiLINKS
NSTA

TOPIC: Succession
GO TO: www.scilinks.org
*sci***LINKS NUMBER:** HSTL470

Primary Succession

Sometimes a small community of living things starts to live in an area that did not previously contain any plants or other organisms. There is no soil in this area, usually just bare rock. Over a very long time, a series of organisms live and die on the rock, and the rock is slowly transformed into soil. This process is called *primary succession.*

1 A slowly retreating glacier exposes bare rock where nothing lives, and primary succession begins.

2 Most primary succession begins with lichens. Acids from the lichens begin breaking the rocks into small particles. These particles mix with the remains of dead lichens to start forming soil. Because lichens are the first organisms to live on the rock, they are called **pioneer species.**

3 After many years, the soil is deep enough for mosses to grow. The mosses eventually replace the lichens. Other tiny organisms, such as insects, also make their home among the lichens and mosses. When they die, their remains add to the soil.

4 Over time, the soil layer thickens, and the moss community is replaced by ferns. The ferns in turn may be replaced by grasses and wild-flowers. Once there is sufficient soil, shrubs and small trees come into the area.

5 After hundreds or even thousands of years, the soil may be deep enough to support a forest.

35

🌐 Multicultural CONNECTION

Lichens contain unique chemicals that help them survive in harsh environments and deter bacteria, herbivores, and competing fungi from consuming the lichens. These chemicals have been used by cultures around the world, mostly as dyes and medicines. Scottish tweedmakers, Navaho weavers, and Chilkat Indians have historically used lichen for dyes. The wolf lichen was used by the Blackfoot tribe as a medicinal tea and to cure skin problems. Chemicals from the lichen genus *Usnea* have been extracted by cultures throughout the world to make medicinal teas and healing creams.

MEETING INDIVIDUAL NEEDS

Learners Having Difficulty
Help students distinguish between primary succession and secondary succession by preparing a table comparing the two. Encourage them to focus on how each type of succession starts, as well as on the organisms that are characteristic of each. Then have students write a sentence summarizing the differences between the two forms of succession. **Sheltered English**

GOING FURTHER

Writing Point out to students that while forest ecosystems are often used as an example to illustrate the process of succession, the process actually occurs in every ecosystem. Encourage students to select an ecosystem to investigate. Have them use library or Internet resources to find out how succession occurs in their chosen ecosystem. Have them prepare brief reports and encourage them to share their findings with the class.

Answer to Self-Check

The main difference between primary and secondary succession is that primary succession begins with the formation of soil. Secondary succession begins on preexisting soil, such as when an existing community is disrupted by a natural disaster or by farming. Pioneer species in primary succession are usually lichens, which begin the formation of soil. Pioneer species in secondary succession are usually seed plants, which germinate and take root in the soil.

Teaching Transparency 71
"Secondary Succession"

Sometimes an existing community is destroyed by a natural disaster, such as fire or flood. Or, a farmer might stop growing crops in an area that had been cleared. In either case, if soil is left intact, the original plant community may regrow through a series of stages called *secondary succession.*

1 The first year after a farmer stops growing crops, or after some other major disturbance, many weeds grow. Crabgrass is usually the most common weed during the first year.

2 By the second year, new weedy plants appear. Their seeds may have blown into the field by the wind, or insects may have carried them. One of the most common weeds during the second year is horseweed.

3 In 5 to 15 years, small pine trees may start growing among the weeds. The pines continue to grow, and, after about 100 years, a forest may form.

4 As older pines die, they may be replaced by hardwoods if the climate can support them.

> ✓ **Self-Check**
>
> Describe the differences between primary succession and secondary succession.
> *(See page 168 to check your answer.)*

36

Homework

Have students identify each of the following examples as either primary succession (p) or secondary succession (s).

- a rock face exposed by a scraping glacier (p)
- a new shoreline in Hawaii created after a lava flow (p)
- a once forested area cleared by a tornado (s)
- an island appearing from the activity of an undersea volcano (p)
- the lawn of an abandoned farmhouse (s)

Where Does It All End? In the early stages of succession only a few species grow in an area. These species grow fast and make many seeds that scatter easily. Because there are only a few species, they are open to invasion by other, longer-lasting species, disease, and other disturbances. In later stages of succession there are usually many more species present. Because of this, there are more pathways available to absorb disturbances. For example, in a mature forest, many species will survive an invasion by insects if these insects prefer to eat only one species of plant.

Eventually, if an area experiences no fires or other disturbances, it will reach a more or less stable stage. Communities change over time even though they are considered to be stable. A stable community may not always be a hardwood forest. Look at **Figure 6.** Why might a stable hardwood forest not develop there? The answer is that the area does not have the kind of climate that will support a stable hardwood forest. The climate in this area supports a desert community.

Figure 6 *This is how a stable community in the Sonoran Desert in Arizona looks in spring.*

Activity

Make a diorama of the stages of primary succession. Use boxes, craft supplies, rocks, twigs, and other materials that you can find at home or at school. You can use one large box showing all the stages of primary succession, or you might use several small boxes, each showing a single stage of primary succession. Label and explain each stage.

TRY at HOME

SECTION REVIEW

1. Define *succession*.

2. Describe succession in an abandoned field.

3. **Applying Concepts** Explain why soil formation is always the first stage of primary succession. Does soil formation stop when trees begin to grow? Why or why not?

internet**connect**

SC*i*LINKS
NSTA

TOPIC: Succession
GO TO: www.scilinks.org
*sci*LINKS NUMBER: HSTL470

Answer to Activity

Dioramas will vary. Be sure students follow the stages described on page 35.

Quiz

1. Describe the main difference between primary and secondary succession. (Primary succession takes place on newly exposed surfaces, where no living things have ever grown before. Secondary succession occurs when an existing community is destroyed and then regrows through a series of stages.)

2. Describe how pioneer species prepare an area for other living things. (Chemicals released by the pioneer species break rock into smaller pieces, and their decaying matter enriches the soil.)

ALTERNATIVE ASSESSMENT

Writing Divide the class into small groups. Challenge each to write a song or play that describes either primary or secondary succession. Direct them to include how succession begins as well as the organisms characteristic of each stage. Encourage students to perform their creations for the class.

PORTFOLIO

37

▼ **Answers to Section Review**

1. Succession is the gradual regrowth or development of a community over time.

2. Sample answer: Succession usually begins with weeds in an abandoned field. After several years, small trees or pine trees begin to grow. After as long as 100 years, the field may grow hardwood trees. Animal life changes with each stage of secondary succession.

3. Soil formation must occur before plants with roots can grow and continues even after a community has developed. Leaves, tree branches, and dead organisms contribute to the soil when they decompose.

Making Models Lab

A Passel o' Pioneers
Teacher's Notes

Time Required

One 45-minute class period and three 5-minute observation periods

Lab Ratings

EASY ———————————→ HARD

TEACHER PREP   🍶🍶
STUDENT SET-UP 🍶🍶
CONCEPT LEVEL 🍶🍶🍶
CLEAN UP   🍶🍶

MATERIALS

The materials suggested for this lab are enough for the entire class. You may want to use a large container, such as a 10 gal aquarium, so each student in the class can contribute a small amount of soil from different areas. Students might be able to bring fishbowls from home. If you have enough fishbowls and sunny windows, students really might enjoy seeing what comes up in their own container. If you do not have a sunny window, the containers can be placed where they will get the most light in the classroom. They need warmth to germinate, and light is necessary after the seedlings emerge from the soil. A local-area field guide might be helpful in identifying plants.

Safety Caution

Remind students to review all safety cautions and icons before beginning this lab activity. Check for any known allergies to molds before having students begin this lab. There may be molds in the soil.

A Passel o' Pioneers

Succession is the progressive replacement of one type of community by another in a single area. The area could be one that has never seen life before and has no soil, such as a cooled lava flow or a rock uncovered by a retreating glacier. In an area where there is no soil, the process is called primary succession. In an area where soil is already there, such as a forest after a fire, the process is called secondary succession. In this exercise, you will build a model of secondary succession using natural soil.

MATERIALS

- large fishbowl
- 500 g of soil from home or schoolyard
- balance
- 250 mL graduated cylinder
- plastic wrap
- water
- protective gloves

Procedure

1 Using a balance, measure 500 g of the soil you brought from home or the schoolyard. Place the soil into the fishbowl. Wet the soil with 250 mL of water. Cover the top of the fishbowl with plastic wrap, and place the fishbowl in a sunny window.
Caution: Do not touch your face, eyes, or mouth during this exercise. Wash your hands when you are finished.

2 For two weeks, watch the soil for any new growth. Describe and draw any new plants you see. Record these and all other observations in your ScienceLog.

3 Name and record as many of these new plants as you can.

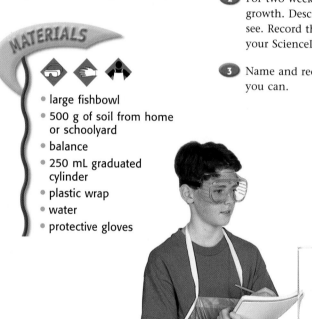

38

 Datasheets for LabBook

Kerry Johnson
Isbell Middle School
Santa Paula, California

Analysis

4 What kinds of plants grew in your model of secondary succession? Were they tree seedlings, grass, or weeds?

5 Were the plants that sprouted in the fishbowl strange or ordinary for your area?

6 Explain how the plants that grew in your model of secondary succession can be called pioneer species.

7 Using your observations, explain how ecological succession worked to maintain equilibrium in your model.

Going Further

Look at each picture on this page. Analyze whether each area, if left alone, would go through primary or secondary succession. You may decide that an area will not go through succession at all. Explain your reasoning.

Going Further

Pond choked with vegetation: A eutrophic pond will eventually fill with sediment. First aquatic plants will grow. When the pond is completely filled, the area will undergo secondary succession.

Bulldozed land: Left abandoned, this area will undergo secondary succession just as abandoned farmland will.

Mount St. Helens volcano: When Mount St. Helens exploded, it caused an enormous mudslide and released tons of ash. Areas smothered by ash underwent primary succession. The entire region is now in various stages of secondary succession.

A pond choked with vegetation

Bulldozed land

Mount St. Helens volcano

39

Answers

4. Plants that sprout should be plants common to the local area. They can be identified using a local-area field guide as soon as dicot seedlings have developed true leaves. Monocots will probably look like grasses.

5. Answers will vary. Students should recognize common plants.

6. All the plants that come up in students' models of succession are pioneers because they are the first plants to grow in an area. Ask students if a tree can be a pioneer. (Yes; if it is one of the first plants to grow in an area. However, pioneers are usually small, fast-growing plants, such as grasses.

7. Answers will vary, but students should recognize that ecological succession allowed their ecosystem to develop toward a state of relative equilibrium.

Chapter Highlights

Chapter Highlights

VOCABULARY DEFINITIONS

SECTION 1

precipitation water in liquid or solid form that moves from the atmosphere to the land and ocean

evaporation the change from a liquid to a vapor

ground water water that is found in underground caverns or in porous rock below the Earth's surface

decomposition the breakdown of dead materials into water and carbon dioxide

combustion the burning of fuel; specifically, the process in which fuel combines with oxygen in a chemical change that produces thermal energy

SECTION 1

Vocabulary

> **precipitation** *(p. 30)*
> **evaporation** *(p. 31)*
> **ground water** *(p. 31)*
> **decomposition** *(p. 32)*
> **combustion** *(p. 32)*

Section Notes

- Materials used by living things continually cycle through ecosystems.
- In the water cycle, water moves through the ocean, atmosphere, land, and living things.
- Precipitation, evaporation, transpiration, and condensation are important processes in the water cycle.

- Water that falls is held in soil or porous rocks as ground water.
- Photosynthesis, respiration, decomposition, and combustion are important steps in the carbon cycle.
- Carbon enters plants from the nonliving environment as carbon dioxide.
- The process of changing nitrogen gas into forms that plants can use is called nitrogen fixation.

Labs

Nitrogen Needs *(p. 130)*

☑ Skills Check

Math Concepts

SAVING WATER Flushing the toilet accounts for almost half the water a person uses in a day. Some toilets use up to 6 gal per flush. More-efficient toilets use about 1.5 gal per flush. How many liters of water can you save using a more-efficient toilet if you flush five times a day?

> 6 gal − 1.5 gal = 4.5 gal
> 4.5 gal × 5 flushes = 22.5 gal
> 1 gal is equal to 3.79 L
> 3.79 L × 22.5 gal = 85.275 L of water saved

Visual Understanding

SOIL FORMATION The formation of soil is part of every stage of primary succession. Look at page 35 to review how soil forms.

Lab and Activity Highlights

A Passel o' Pioneers PG 38

Nitrogen Needs PG 130

Datasheets for LabBook
(blackline masters for these labs)

SECTION 2

succession (ecological succession) the gradual regrowth or development of a community of organisms over time

pioneer species the first organisms to grow in an area undergoing ecological succession; usually lichens in primary succession and weedy plants in secondary succession

Vocabulary

succession *(p. 34)*

pioneer species *(p. 35)*

Section Notes

- Ecological succession is the gradual development of communities over time. Often a series of stages is observed during succession.

- Primary succession occurs in an area that was not previously inhabited by living things; no soil is present.

- Secondary succession occurs in an area where an earlier community was disturbed by fire, landslides, floods, or plowing for crops; soil is present.

 Vocabulary Review Worksheet

 Blackline masters of these Chapter Highlights can be found in the **Study Guide.**

---

internetconnect

 GO TO: go.hrw.com

 N S T A **GO TO:** www.scilinks.org

Visit the **HRW** Web site for a variety of learning tools related to this chapter. Just type in the keyword:

KEYWORD: HSTCYC

Visit the **National Science Teachers Association** on-line Web site for Internet resources related to this chapter. Just type in the *sci*LINKS number for more information about the topic:

TOPIC: The Water Cycle	*sci*LINKS NUMBER: HSTL455
TOPIC: The Carbon Cycle	*sci*LINKS NUMBER: HSTL460
TOPIC: The Nitrogen Cycle	*sci*LINKS NUMBER: HSTL465
TOPIC: Succession	*sci*LINKS NUMBER: HSTL470

41

Lab and Activity Highlights

LabBank

 Long-Term Projects & Research Ideas,
Smokey Says . . .

Chapter Review

USING VOCABULARY

To complete the following sentences, choose the correct term from each pair of terms listed below:

1. During ___?___, water moves from the atmosphere to the land and ocean. (*evaporation* or *precipitation*)

2. All biological molecules contain ___?___. (*carbon* or *carbon dioxide*)

3. The combustion of coal, oil, and natural gas is part of the ___?___. (*nitrogen cycle* or *carbon cycle*)

4. The development of a community on bare, exposed rock is an example of ___?___. (*primary succession* or *secondary succession*)

5. The recovery of Yellowstone National Park following the fires of 1988 is an example of ___?___. (*primary succession* or *secondary succession*)

UNDERSTANDING CONCEPTS

Multiple Choice

6. Water changes from a liquid to a vapor during
   a. precipitation.
   b. respiration.
   c. evaporation.
   d. decomposition.

7. The process of burning fuel, such as oil and coal, is
   a. combustion.
   b. respiration.
   c. decomposition.
   d. photosynthesis.

8. One of the most common plants in a recently abandoned farm field is
   a. horseweed.
   b. young pine trees.
   c. young oak and hickory trees.
   d. crabgrass.

9. Which of the following statements about ground water is true?
   a. It stays underground for a few days.
   b. It is stored in underground caverns or porous rock.
   c. It is salty like ocean water.
   d. It never reenters the water cycle.

10. Which of the following processes produces carbon dioxide?
    a. decomposition
    b. respiration
    c. combustion
    d. all of the above

11. During nitrogen fixation, nitrogen gas is converted into a form that ___?___ can use.
    a. plants
    b. animals
    c. fungi
    d. all of the above

12. Bacteria are essential to
    a. combustion.
    b. photosynthesis.
    c. nitrogen fixation.
    d. evaporation.

13. The pioneer species on bare rock are usually
    a. ferns.
    b. pine trees.
    c. mosses.
    d. lichens.

Short Answer

14. Is snow a part of the water cycle? Why or why not?

15. Can a single scientist observe all of the stages of secondary succession on an abandoned field? Explain your answer.

42

Concept Mapping Transparency 19

Blackline masters of this Chapter Review can be found in the **Study Guide.**

Concept Mapping

16. Use the following terms to create a concept map: abandoned farmland, lichens, bare rock, soil formation, horseweed, succession, forest fire, primary succession, secondary succession, pioneer species.

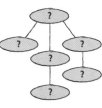

CRITICAL THINKING AND PROBLEM SOLVING

Write one or two sentences to answer the following questions:

17. Explain how living things would be affected if the water on our planet suddenly stopped evaporating.

18. How would living things be affected if there were no decomposers to cycle carbon back to the atmosphere?

19. Explain how living things would be affected if the bacteria responsible for nitrogen fixation were to die.

20. Describe why a lawn doesn't go through succession.

MATH IN SCIENCE

In 1996, 129 million metric tons of fertilizer were used world-wide. Use the following information to answer items 21, 22, and 23: 1996 world population = 5.7 billion; 1 metric ton = 1,000 kg; 1 kg = 2.2 lb.

21. Write out the number corresponding to 5.7 billion. How many zeros are in the number?

22. How many kilograms of fertilizer were used per person in 1996?

23. How many pounds of fertilizer were used per person?

INTERPRETING GRAPHICS

The following graph illustrates the concentration of carbon dioxide in the atmosphere from 1958 to 1994:

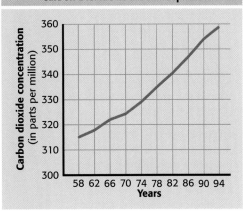

Carbon Dioxide in the Atmosphere

24. What was the concentration of carbon dioxide in parts per million in 1960? in 1994?

25. Is the concentration of carbon dioxide increasing or decreasing? Explain.

26. If the level of carbon dioxide continues to change at the same steady rate, what might be the concentration in 2010?

Reading Check-up

Take a minute to review your answers to the Pre-Reading Questions found at the bottom of page 28. Have your answers changed? If necessary, revise your answers based on what you have learned since you began this chapter.

Concept Mapping

16. An answer to this exercise can be found at the front of this book.

CRITICAL THINKING AND PROBLEM SOLVING

17. Sample answer: The Earth would soon be covered with water. The water cycle would be interrupted. It would no longer rain. Land organisms would probably go extinct.

18. Sample answer: The carbon cycle would be interrupted. Dead organisms would remain unchanged. Plants could not perform photosynthesis, and there would be no food.

19. Sample answer: If the nitrogen cycle were interrupted, nitrogen might be locked in the atmosphere in an unusable form. Plants might not be able to obtain enough of the nitrogen needed for life. Living things might eventually run out of food.

20. A lawn does not go through the stages of succession because it is constantly being "disturbed" by mowing and other activities.

MATH IN SCIENCE

21. 5,700,000,000; 8 zeros
22. 22.6 kg
23. 22.6 kg × 2.2 lb = 49.7 lb

INTERPRETING GRAPHICS

24. 1960, 317 ppm; 1994, 359 ppm
25. Carbon dioxide in the atmosphere is showing a steady increase.
26. By extending the graph and drawing a straight line from coordinates (58, 315) through (92, 356), the concentration would be 378 ppm in 2010. There are other ways to estimate this figure.

Background

For massive plumes of flame and hot air to develop, a fire must have a wealth of fuel, such as very dry twigs and leaves, and there must be warm, rising currents of air. But wind is by far the most important factor in the creation of plumes. A strong crosswind can inhibit the formation of a plume.

Forest fires break out thousands of times a year around the world. In 1985, a total of 81,622 fires destroyed about 1.2 million hectares of forest in the United States alone. In 1871, fires in Wisconsin and Michigan killed 1,500 people and burned 1.7 million hectares of land—an area more than five times the size of Rhode Island!

Another type of interaction between fires and the weather occurs in confined regions, such as valleys. As the upper atmosphere cools after sunset, a temperature inversion occurs, causing warm air to be trapped beneath the denser, cooler air above it. A fire in the valley gradually uses up all of the oxygen in the trapped warm air and eventually starts to smolder, heating up the unburned material surrounding it. When the sun rises and heats the overlying cool air, the oxygen-depleted air escapes and is replaced with oxygen-rich air. The result can be an explosively renewed forest fire.

WEIRD SCIENCE

WEATHER FROM FIRE

As a wildfire burned near Santa Barbara, California, in 1993, huge storm clouds formed overhead. Fiery whirlwinds danced over the ground. The fire not only was destroying everything in its path—it was also creating its own weather!

Fire-Made Clouds

Hot air rising from a forest fire can create tremendous updrafts. Surrounding air rushes in underneath the rising air, stirring up columns of ash, smoke, hot air, and noxious gases. Cool, dry air normally sinks down and stops these columns from developing any further. But if the conditions are just right, a surprising thing happens.

If the upper atmosphere contains warm, moist air, the moisture begins to condense on the ash and smoke. These droplets can develop into clouds. As the clouds grow, the droplets begin to collide and combine until they are heavy enough to fall as rain. The result is an isolated rainstorm, complete with thunder and lightning.

▲ *This towering whirlwind is lifting burning debris from a forest fire in Idaho.*

Whirlwinds of Fire

Forest fires can also create whirlwinds. These small, tornado-like funnels can be extremely dangerous. Whirlwinds are similar to dust devils that dance across desert sands. Their circular motion is created by an updraft that is forced to turn after striking an obstacle, such as a cliff or hill. Whirlwinds move across the ground at 8–11 km/h, sometimes growing up to 120 m high and 15 m wide.

Most whirlwinds last less than a minute, but they can cause some big problems. Fire-fighters caught in the path of whirlwinds have been severely injured and even killed. Also, if a whirlwind is hot enough, it can suck up tremendous amounts of air. The resulting updraft can pull burning debris up through the whirlwind. In some cases, burning trees have been uprooted and shot into the air. When the debris lands, it often starts new fires hundreds of meters away.

Think About It

▶ Fires are a natural part of the growth of a forest. For example, some tree seeds are released only under the extreme temperatures of a fire. Some scientists believe that forest fires should be allowed to run their natural course. Others argue that forest fires cause too much damage and should be extinguished as soon as possible. Do some additional research and then decide what you think.

44

EYE ON THE ENVIRONMENT

The Mysterious Dead Zone

Every summer, millions of fish are killed in an area in the Gulf of Mexico called a hypoxia region. Hypoxia is a condition that occurs when there is an unusually low level of oxygen in the water. The area is often referred to as the "dead zone" because almost every fish and crustacean in the area dies. In 1995, this zone covered more than 18,000 km^2, and almost 1 million fish were killed in a single week. Why does this happen? Can it be stopped?

▲ *The Gulf of Mexico hypoxia region grew to the size of New Jersey in 1995.*

What's Going On?

When the oxygen levels in water drop drastically, the fish die. In the Gulf of Mexico hypoxia region, the water contains unusually high amounts of nitrogen and phosphorus. The nitrogen and phosphorus act as nutrients for the growth of algae. When the algae die, their bodies are decomposed by a large number of oxygen-consuming bacteria.

Scientists think the excess nitrogen and phosphate is from animal waste and runoff from farms and developments. The pollution may also be caused in part by the overfertilization of crops. The extra fertilizer runs into the rivers, which empty into the Gulf Coast.

Ecosystem Models Suggest Solutions

All along the Gulf Coast, marine scientists and Earth scientists are trying to find methods to reduce or eliminate the "dead zone." They have made physical and computer models of the Mississippi River ecosystem that have accurately predicted the data that has since been collected. The scientists have changed the models to see what happens. For example, wetlands are one of nature's best filters. They take up a lot of the chemicals present in water. Scientists predict that adding wetlands to the Mississippi River watershed could reduce the chemicals reaching the Gulf of Mexico. Although scientific models support this hypothesis, they also indicate that adding wetlands to the Mississippi River watershed would not be enough to completely prevent the "dead zone."

Find Some Solutions

▶ The Gulf of Mexico is not the only place that suffers from a hypoxia region. Research other bodies of water to find out how widespread the problem is. Have scientists found ways to reduce or eliminate the hypoxia regions elsewhere? How could this information be used to improve the situation in the Gulf of Mexico?

Background

The first law of ecology states that everything is connected to everything else. Clearly, the Dead Zone is an example of how human activity exacerbates natural seasonal variations in dissolved oxygen. Before the Mississippi watershed was controlled, regular flooding spread nutrients on the sides of the rivers, while numerous wetlands acted as an additional nutrient control. Evidence for the chemical runoff connection to the zone was demonstrated during the flood of 1993, when the zone doubled in size.

Answer to Find Some Solutions

Answers will vary. In Long Island Sound, for example, the plan for managing hypoxia will require a great deal of money and the coordinated efforts of many states and people. The nitrogen loads deposited into the sound are being identified, monitored, and gradually reduced over time.

45

Chapter Organizer

CHAPTER ORGANIZATION	TIME MINUTES	OBJECTIVES	LABS, INVESTIGATIONS, AND DEMONSTRATIONS
Chapter Opener pp. 46–47	45	National Standards: SAI 1, ST 2, SPSP 2, 5, HNS 1, LS 3a, 3c, 4a, 4b	**Start-Up Activity,** A Mini-Ecosystem, p. 47
Section 1 Land Ecosystems	90	▶ Define *biome*. ▶ Describe three different forest biomes. ▶ Distinguish between temperate grasslands and savannas. ▶ Describe the importance of permafrost to the arctic tundra biome. UCP 4, 5, SAI 1, SPSP 2, 3, LS 1a, 3a, 3c, 3d, 4b–4d, 5a, 5b; Labs UCP 2, SAI 1, ST 1	**Design Your Own,** Life in the Desert, p. 132 **Datasheets for LabBook,** Life in the Desert
Section 2 Marine Ecosystems	90	▶ Distinguish between the different areas of the ocean. ▶ Explain the importance of plankton in marine ecosystems. ▶ Describe coral reefs and intertidal areas. UCP 2, 3, SAI 2, SPSP 2, HNS 2, LS 1a, 3d, 4a–4d; Labs SAI 1, LS 1a	**Interactive Explorations CD-ROM,** Sea Sick *A **Worksheet** is also available in the **Interactive Explorations Teacher's Edition.*** **Design Your Own,** Discovering Mini-Ecosystems, p. 133 **Datasheets for LabBook,** Discovering Mini-Ecosystems
Section 3 Freshwater Ecosystems	90	▶ List the characteristics of rivers and streams. ▶ Describe the littoral zone of a pond. ▶ Distinguish between two types of wetlands. UCP 2–4, LS 1a, 3d, 4a–4d; Labs SAI 1, SPSP 2	**QuickLab,** Pond Food Connections, p. 61 **Skill Builder,** Too Much of a Good Thing? p. 64 **Datasheets for LabBook,** Too Much of a Good Thing? **EcoLabs & Field Activities,** Biome Adventure Travel **Long-Term Projects & Research Ideas,** Tropical Medicine

*See page **T23** for a complete correlation of this book with the*

NATIONAL SCIENCE EDUCATION STANDARDS.

TECHNOLOGY RESOURCES

 Guided Reading Audio CD English or Spanish, Chapter 3

 One-Stop Planner CD-ROM with Test Generator

 Interactive Explorations CD-ROM CD 2, Exploration 2, Sea Sick

 CNN Multicultural Connections, Saving Pacific Sea Horses, Segment 9

Eye on the Environment, Biosphere Pioneers, Segment 6

 Science Discovery Videodiscs Image and Activity Bank with Lesson Plans: Tragedies in the Commons

Chapter 3 • The Earth's Ecosystems

CLASSROOM WORKSHEETS, TRANSPARENCIES, AND RESOURCES	SCIENCE INTEGRATION AND CONNECTIONS	REVIEW AND ASSESSMENT
Directed Reading Worksheet **Science Puzzlers, Twisters & Teasers**	**Careers:** Ecologist—Alfonso Alonso-Mejía, p. 71	
Transparency 72, Earth's Biomes **Directed Reading Worksheet,** Section 1 **Math Skills for Science Worksheet,** Subtraction Review **Transparency 73,** Coniferous Forest Biome **Transparency 74,** A Tropical Rain Forest Biome **Math Skills for Science Worksheet,** Rain-Forest Math **Transparency 177,** An Example of the Rain Shadow Effect **Reinforcement Worksheet,** Know Your Biomes	**Cross-Disciplinary Focus,** p. 49 in ATE **Math and More,** p. 50 in ATE **Real-World Connection,** p. 51 in ATE **Multicultural Connection,** p. 51 in ATE **Multicultural Connection,** p. 52 in ATE **Connect to Earth Science,** p. 52 in ATE **MathBreak,** Rainfall, p. 54	**Self-Check,** p. 53 **Section Review,** p. 54 **Quiz,** p. 54 in ATE **Alternative Assessment,** p. 54 in ATE
Directed Reading Worksheet, Section 2 **Science Skills Worksheet,** Being Flexible	**Multicultural Connection,** p. 56 in ATE **Real-World Connection,** p. 57 in ATE **Connect to Physical Science,** p. 57 in ATE **Across the Sciences:** Ocean Vents, p. 70	**Self-Check,** p. 58 **Section Review,** p. 59 **Quiz,** p. 59 in ATE **Alternative Assessment,** p. 59 in ATE
Transparency 75, River Features **Directed Reading Worksheet,** Section 3 **Transparency 76,** Lake Zones **Critical Thinking Worksheet,** Risky Development?	**Math and More,** p. 61 in ATE **Connect to Physical Science,** p. 61 in ATE **Apply,** p. 62	**Section Review,** p. 63 **Quiz,** p. 63 in ATE **Alternative Assessment,** p. 63 in ATE

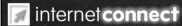 internet**connect**

 go. hrw .com **Holt, Rinehart and Winston On-line Resources**
go.hrw.com

For worksheets and other teaching aids related to this chapter, visit the HRW Web site and type in the keyword: **HSTECO**

 SC*LINKS* NSTA **National Science Teachers Association**
www.scilinks.org

Encourage students to use the *sci*LINKS numbers listed in the internet connect boxes to access information and resources on the **NSTA** Web site.

END-OF-CHAPTER REVIEW AND ASSESSMENT

Chapter Review in Study Guide
Vocabulary and Notes in Study Guide
Chapter Tests with Performance-Based Assessment, Chapter 3 Test
Chapter Tests with Performance-Based Assessment, Performance-Based Assessment 3
Concept Mapping Transparency 20

Chapter Resources & Worksheets

Visual Resources

TEACHING TRANSPARENCIES

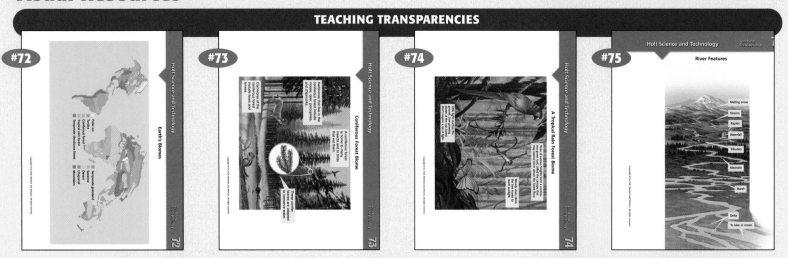

#72 Earth's Biomes — Holt Science and Technology

#73 Coniferous Forest Biome — Holt Science and Technology

#74 A Tropical Rain Forest Biome — Holt Science and Technology

#75 Holt Science and Technology — River Features

TEACHING TRANSPARENCIES

#76 Lake Zones — Holt Science and Technology

#177 An Example of the Rain Shadow Effect — Holt Science and Technology

LINK TO EARTH SCIENCE

CONCEPT MAPPING TRANSPARENCY

#20 Holt Science and Technology — Concept Mapping Transparency 20

The Earth's Ecosystems

Use the following terms to complete the concept map below: temperature, biomes, land biomes, water biomes, marine, rainfall, tundra, coastal areas, grasslands

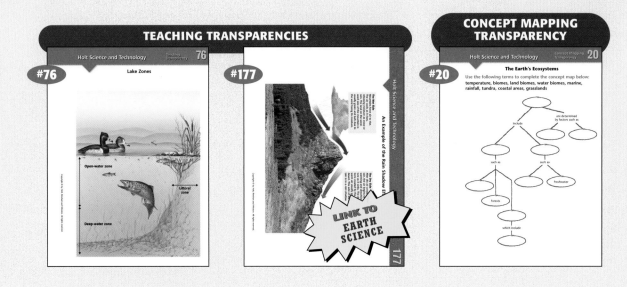

Meeting Individual Needs

DIRECTED READING

#3 DIRECTED READING WORKSHEET
The Earth's Ecosystems

Chapter Introduction

As you begin this chapter, answer the following.

1. Read the title of the chapter. List three things that you already know about this subject.

2. Write two questions about this subject that you would like answered by the time you finish this chapter.

3. How does the title of the Start-Up Activity relate to the subject of the chapter?

Section 1: Land Ecosystems (p. 60)

4. _____ factors are features of an environment that are not alive, such as temperature and rainfall.

REINFORCEMENT & VOCABULARY REVIEW

#3 REINFORCEMENT WORKSHEET
Know Your Biomes

Complete this worksheet after you have finished reading Chapter 17, Section 1.

1. Using the Temperature & rainfall column as a guide, label the type of biomes listed with the following terms: *desert, tropical rain forest, arctic tundra, coniferous forest, temperate grassland, savanna,* and *temperate deciduous forest.*

2. Use the examples and characteristics given in the box on the next page to fill in the appropriate blanks.

Type of biome	Temperature & rainfall	Examples & characteristics
	summer: 38°C	jackrabbit
	winter: 7°C	
	rain: less than 25 cm per year	
	dry season: 34°C	has scattered clumps of trees
	wet season: 16°C	
	rain: 150 cm per year	
	daytime: 34°C	the most biologically diverse biome
	nighttime: 20°C	
	rain: up to 400 cm per year	
	summer: 28°C	woody shrubs beneath tree layer
	winter: 6°C	
	rain: 75–125 cm per year	

#3 VOCABULARY REVIEW WORKSHEET
Eco-Puzzle

After you finish Chapter 17, give this puzzle a try! In the space provided, write the term described by the clue. Then find those words in the puzzle. Terms can be hidden in the puzzle vertically, horizontally, or diagonally.

1. a biome in the far north _____
2. a tree that produces seeds in a cone _____
3. soil that is always frozen _____
4. a biome that receives less than 25 cm of rain a year _____
5. the shore of a large body of water is this type of zone _____
6. a treeless wetland ecosystem _____
7. microscopic photosynthetic organisms in the sea _____
8. geographic area characterized by certain types of plants and animals _____
9. trees that lose their leaves in winter are _____
10. a wetland ecosystem with trees _____
11. an algae that forms "rafts" in the Atlantic _____
12. an area where fresh and salty waters mix _____
13. land where the water level is near or above the surface of the ground _____
14. microscopic consumers in the sea _____
15. a tropical grassland with clumps of trees _____

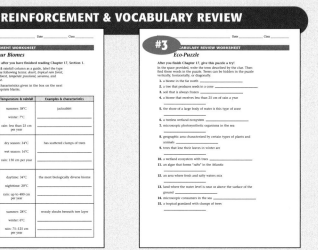

SCIENCE PUZZLERS, TWISTERS & TEASERS

#3 SCIENCE PUZZLERS, TWISTERS & TEASERS
The Earth's Ecosystems

Eco Tourism

1. You've wandered into a travel agency that offers trips to any ecosystem worth visiting. Inside, you pick up a brochure. Identify the ecosystems described below.

   a. You'll love the lush growth and busy chatter of countless plants and animals. A must-see for the bug fan in you, and a great winter getaway!

   b. Many have described it as a "sea of grass" peppered with wildflowers. Said one visitor, "I hardly even missed the trees!"

   c. Fall is paradise here! The colors peak in September and October, so make your reservations before the last leaf drops!

   d. Lakes, ponds, and wide open spaces! Come in the summer to see the birds, and remember your boots—it gets soggy!

   e. The breeze through the trees will bring the fresh scent of pine to your tent as you prepare for the morning's hike.

   f. Work on your tan while enjoying the beautiful, stark scenery. And don't forget to bring plenty of water along.

Gently Down the Stream

2. Below are what's left of three words that have been mixed up in a raging river. Each word came out of the water as two or more different words. Rearrange the letters to find the original words, which are all features of a river.

   a. BURY IT RAT _____

   b. LET FAR LAW _____

   c. READ MEN _____

Chapter 3 • The Earth's Ecosystems

Review & Assessment

STUDY GUIDE

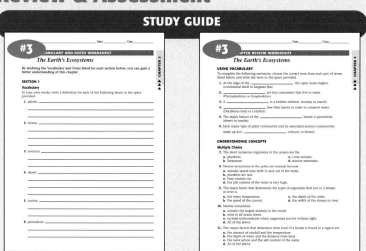

#3 VOCABULARY AND NOTES WORKSHEET
The Earth's Ecosystems

By studying the Vocabulary and Notes listed for each section below, you can gain a better understanding of this chapter.

SECTION 1
Vocabulary
In your own words, write a definition for each of the following terms in the space provided.

1. abiotic
2. biome
3. savanna
4. desert
5. tundra
6. permafrost

#3 CHAPTER REVIEW WORKSHEET
The Earth's Ecosystems

USING VOCABULARY
To complete the following sentences, choose the correct term from each pair of terms listed below, and write the term in the space provided.

1. At the edge of the _____, the open ocean begins. (continental shelf or Sargasso Sea)
2. _____ are tiny consumers that live in water. (Phytoplankton or Zooplankton)
3. A _____ is a treeless wetland. (swamp or marsh)
4. _____ (Deciduous trees or Conifers) lose their leaves in the fall to conserve water.
5. The major feature of the _____ biome is permafrost. (desert or tundra)
6. Each major type of plant community and its associated animal communities make up a(n) _____. (estuary or biome)

UNDERSTANDING CONCEPTS
Multiple Choice

7. The most numerous organisms in the oceans are the
   a. plankton.   c. coral animals.
   b. Sargassum.   d. marine mammals.
8. Marine ecosystems at the poles are unusual because
   a. animals spend time both in and out of the water.
   b. plankton are rare.
   c. they contain ice.
   d. the salt content of the water is very high.
9. The major factor that determines the types of organisms that live in a stream or river is
   a. the water temperature.   c. the depth of the water.
   b. the speed of the current. d. the width of the stream or river.
10. Marine ecosystems
    a. contain the largest animals in the world.
    b. exist in all ocean zones.
    c. include environments where organisms survive without light.
    d. All of the above
11. Two major factors that determine what kind of a biome is found in a region are
    a. the amount of rainfall and the temperature.
    b. the depth of water and the distance from land.
    c. the wave action and the salt content of the water.
    d. All of the above

CHAPTER TESTS WITH PERFORMANCE-BASED ASSESSMENT

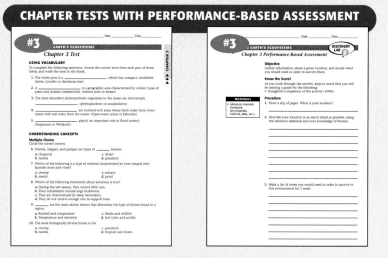

#3 THE EARTH'S ECOSYSTEMS
Chapter 3 Test

USING VOCABULARY
To complete the following sentences, choose the correct term from each pair of terms listed, and write the term in the blank.

1. The white pine is a _____, which has compact, needlelike leaves. (conifer or deciduous tree)
2. A _____ is a geographic area characterized by certain types of plant and animal communities. (littoral zone or biome)
3. The most abundant photosynthetic organisms in the ocean are microscopic _____. (phytoplankton or zooplankton)
4. _____ are nutrient-rich areas where fresh water from rivers mixes with salt water from the ocean. (Open-water zones or Estuaries)
5. _____ play(s) an important role in flood control. (Sargassum or Wetlands)

UNDERSTANDING CONCEPTS
Multiple Choice
Circle the correct answer.

6. Prairies, steppes, and pampas are types of _____ biomes.
   a. chaparral   c. desert
   b. tundra   d. grassland
7. Which of the following is a type of wetland characterized by trees draped with Spanish moss and vines?
   a. swamp   c. estuary
   b. marsh   d. pond
8. Which of the following statements about savannas is true?
   a. During the wet season, they receive little rain.
   b. Their inhabitants include large herbivores.
   c. They are characterized by steep mountains.
   d. They do not receive enough rain to support trees.
9. _____ are the main abiotic factors that determine the type of biome found in a region.
   a. Rainfall and temperature   c. Shade and wildlife
   b. Temperature and elevation d. Soil color and acidity
10. The most biologically diverse biome is the
    a. swamp   c. grassland.
    b. tundra.   d. tropical rain forest.

#3 THE EARTH'S ECOSYSTEMS
Chapter 3 Performance-Based Assessment

Objective
Gather information about a given location, and decide what you would need in order to survive there.

Know the Score!
As you work through the activity, keep in mind that you will be earning a grade for the following:
• thoughtful completion of the activity (100%)

MATERIALS
• reference materials (textbook, encyclopedia, internet, atlas, etc.)

Procedure
1. Draw a slip of paper. What is your location?
2. Describe your location in as much detail as possible, using the reference materials and your knowledge of biomes.
3. Make a list of items you would need in order to survive in this environment for 1 week.

Lab Worksheets

ECOLABS & FIELD ACTIVITIES

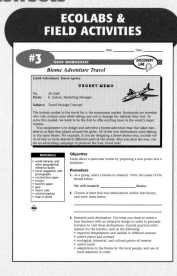

#3 STUDENT WORKSHEET
Biome Adventure Travel

Earth Adventures Travel Agency

URGENT MEMO

To: All Staff
From: E. Concos, Marketing Manager

Subject: Travel Package Concept

The hottest market in the travel biz is the ecotourism market. Ecotourists are travelers who visit natural areas while taking care not to damage the habitats they visit. To serve this market, we want to be the first to offer exciting tours to the world's major biomes.

Your assignment is to design and advertise a biome-adventure tour that takes travelers to at least four places around the globe. All of the tour destinations must belong to the same biome. For example, if you are designing a desert-biome tour, include visits to four or more deserts in different parts of the world. After you plan the tour, create an advertising campaign to promote the tour. Good luck!

MATERIALS
• world almanac and other geographical reference books
• travel magazines with photographs
• construction paper
• scissors
• butcher paper
• glue
• metric ruler
• colored markers
• map or globe

Objective
Learn about a particular biome by preparing a travel poster and a brochure.

Procedure
1. As a group, select a biome to research. Write the name of the biome below.

   We will research _____ biome.

2. Choose at least four tour destinations within that biome, and write them below.

3. Research each destination. Use what you learn to create a tour brochure with an attractive design in order to persuade travelers to visit these destinations. Include practical information for the traveler, such as the following:
   • expected temperatures and rainfall in different seasons
   • native plants and animals
   • ecological, historical, and cultural points of interest
   • native foods
   • adaptations to the biome by the local people, and use of local materials in crafts

LONG-TERM PROJECTS & RESEARCH IDEAS

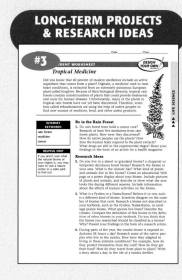

#3 STUDENT WORKSHEET
Tropical Medicine

Did you know that 40 percent of modern medicines include an active ingredient that comes from a plant? Digitalis, a medicine used to treat heart conditions, is extracted from an extremely poisonous European plant called foxglove. Because of their biological diversity, tropical rain forests contain untold numbers of plants that could provide treatments and cures for human diseases. Unfortunately, many of the plants in tropical rain forests have not yet been discovered. Therefore, scientists called ethnobotanists are using the help of native peoples to find new sources of medicine, food, and other useful products.

INTERNET KEYWORDS
rain forest
medicine
cancer

HELPFUL HINT
If you aren't sure what the natural biome of your region is, you may want to visit a nature center or a state park to find out.

Rx in the Rain Forest
1. Do rain forest trees hold a cancer cure? Research at least five medicines from rain forest plants. How were they discovered? How do native peoples use the plants? How does the human body respond to the plant extracts? What drugs are still in the experimental stages? Share your findings in the form of an article for a health magazine.

Research Ideas
2. Do you live in a desert or grassland biome? A chaparral or temperate deciduous forest biome? Research the biome in your area. What is the climate like? What kind of plants and animals live in the biome? Create an educational Web page or a poster display about your biome. Include pictures of plants and animals, and describe or show what the area looks like during different seasons. Include information about the effects of human activities on the biome.

3. What is a Fynbos or a Nama-Karoo? Believe it or not, each is a different kind of biome. Scientists disagree on the number of biomes that exist. Research a biome not described in your textbook, such as the Fynbos, Nama-Karoo, or sandsage prairie biome. What species live there? Describe the climate. Compare the definition of this biome to the definition of other biomes in your textbook. Do you think that the biome you researched should be classified as a biome? Why? Present your findings in the form of a research article.

4. During parts of the year, the tundra biome is exposed to darkness 24 hours a day! Research some of the native peoples who live in the tundra. How have they adapted to living in these extreme conditions? For example, how do they protect themselves from the cold? How do they get their food? How do they travel from place to place? Write a story about a day in the life of a tundra dweller.

DATASHEETS FOR LABBOOK

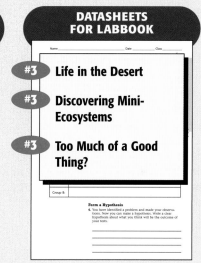

#3 Life in the Desert

#3 Discovering Mini-Ecosystems

#3 Too Much of a Good Thing?

Form a Hypothesis
4. You have identified a problem and made your observations. Now you can make a hypothesis. Write a clear hypothesis about what you think will be the outcome of your tests.

Applications & Extensions

CRITICAL THINKING & PROBLEM SOLVING

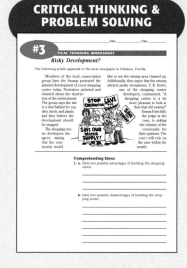

#3 CRITICAL THINKING WORKSHEET
Risky Development?

The following article appeared in the local newspaper in Frikshon, Florida.

Members of the local conservation group Save the Swamp protested the planned development of a new shopping center today. Protesters picketed and chanted about the destruction of the environment. The group says the site is a vital habitat for reptiles, birds, and plants, and they believe the development should be stopped.

The shopping center developers disagree, stating that the community would

like to see the swamp area cleaned up. Additionally, they argue that the swamp attracts pesky mosquitoes. T. R. Down, one of the shopping center developers, commented, "A shopping center is a lot more pleasant to look at than that old swamp!" Emma Fairchild, the judge in the case, is asking the citizens of the community for their opinions. The court will rule on the case within the month.

Comprehending Ideas
1. a. State two possible advantages of building the shopping center.

   b. State two possible disadvantages of building the shopping center.

MULTICULTURAL CONNECTIONS

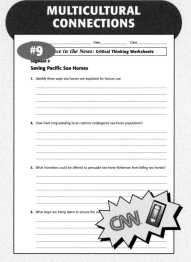

#9 Science in the News: Critical Thinking Worksheets

Segment 9
Saving Pacific Sea Horses

1. Identify three ways sea horses are exploited for human use.

2. How have long-standing local customs endangered sea-horse populations?

3. What incentives could be offered to persuade sea-horse fisherman from killing sea horses?

4. What steps are being taken to ensure the su...

EYE ON THE ENVIRONMENT

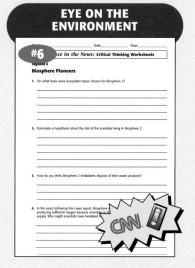

#6 Science in the News: Critical Thinking Worksheets

Segment 6
Biosphere Pioneers

1. On what basis were ecosystem types chosen for Biosphere 2?

2. Formulate a hypothesis about the diet of the scientists living in Biosphere 2.

3. How do you think Biosphere 2 inhabitants dispose of their waste products?

4. In the years following this news report, Biosphere ... producing sufficient oxygen because microbes in th... supply. Why might scientists have hesitated t...

INTERACTIVE EXPLORATIONS

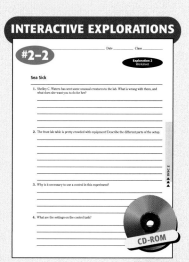

#2-2 Exploration 2 Worksheet

Sea Sick

1. Shelley C. Waters has sent some unusual creatures to the lab. What is wrong with them, and what does she want you to do for her?

2. The front lab table is pretty crowded with equipment! Describe the different parts of the setup.

3. Why is it necessary to use a control in this experiment?

4. What are the settings on the control tank?

CD-ROM

Land Ecosystems

▶ The Biosphere

All parts of Earth that are inhabited by organisms make up the biosphere. The biosphere is a relatively thin layer encircling the planet.

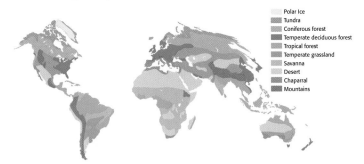

Polar Ice
Tundra
Coniferous forest
Temperate deciduous forest
Tropical forest
Temperate grassland
Savanna
Desert
Chaparral
Mountains

- The Earth can be divided into gas, liquid, and solid parts. The atmosphere is the layer of gases that envelops Earth. The hydrosphere is the portion of Earth's surface that is covered by water. The lithosphere is the soil and rock on Earth's surface.

- Life began in the water and remained in the water for billions of years before migrating to land. Today, aquatic habitats continue to dominate the biosphere.

▶ Biotic and Abiotic Factors

An ecosystem encompasses all of the biotic and abiotic factors in a particular area. Ecology is the scientific study of how organisms interact with one another and with the abiotic factors in their environment. Biotic factors include all organisms—plants, animals, protists, fungi, and bacteria. Abiotic factors include temperature, water, sunlight, wind, rocks, soil, and nutrients.

- Almost all ecosystems are driven by energy from the sun. Thus, the amount of sunlight a region receives has a large effect on the number of producers and consumers that can be supported in the ecosystem.

- Just as chemical reactions are limited by a limiting reagent, populations within ecosystems are limited by limiting factors. The lack of any single abiotic factor can prevent the survival of a population. Limiting factors include the amount of light received, water, temperature, and nutrients in the soil.

IS THAT A FACT!

- ☛ One hectare of fertile soil can contain 6 million earthworms and over 2 billion bacteria!

▶ Tropical Rain Forests

Students may think that tropical rain forests are jungles. In fact, a jungle is an area of dense undergrowth within a tropical rain forest. Jungles grow in areas that receive large amounts of sunlight and are near rivers.

- The plants in a tropical rain forest can be divided into three layers. Some trees tower above the forest's canopy. A continuous layer of vegetation forms the upper canopy. Below the canopy is a mid-tree level and a shrubby understory. On the ground are grasses and ferns.

IS THAT A FACT!

- ☛ Tropical rain forests cover about 7 percent of Earth's surface. This small portion of Earth contains more than 50 percent of the species that inhabit the planet.

▶ Tundra

A tundra area may receive as little rainfall as a desert. But the soil in a tundra region remains wet due to permafrost, low temperatures, and the low rate of evaporation.

IS THAT A FACT!

- ☛ Antarctica has been accumulating ice for more than 25 million years. It contains about 90 percent of Earth's ice and about 70 percent of Earth's fresh water.

- ☛ In summer, when ice begins to melt and break off into icebergs, Antarctica shrinks. In winter, Antarctica expands to twice its summer size.

SECTION 2

Marine Ecosystems

▶ Oceans

The Earth's oceans include the Pacific Ocean, the Atlantic Ocean, the Indian Ocean, and the Arctic Ocean. Although these oceans have different names, they are all connected.

- Because the oceans are all connected, a change in one marine environment may eventually affect other marine environments.

IS THAT A FACT!

➤ Along Australia's northeastern coast is the Great Barrier Reef. It is the largest and most diverse reef system in the world.

▶ Underwater Exploration

New technology for remote-operated vehicles has broadened scientists' ability to explore ocean depths. Using vehicles equipped with cameras, mechanical arms, and remote sensors, scientists have discovered a watery new world that includes deep-sea animals, underwater volcanoes, thermal vents, and entire ecosystems that do not directly depend on light or photosynthesis for energy.

- Early diving suits consisted of a hard helmet, a canvas-and-rubber tunic, leather boots with lead-weighted soles, and additional 13 kg weights. Each boot weighed about 8 kg. Such a suit was called a standard diving suit and was invented by Augustus Siebe in the 1830s.

SECTION 3

Freshwater Ecosystems

▶ Wetlands

In the past, wetlands were underappreciated and even considered wastelands. People viewed wetlands as places that should be drained or filled in so that the land could be used for housing or other urban development. Because wetlands are a breeding ground for mosquitoes, they were also considered a health problem.

- Two types of wetlands are introduced in the text: the marsh and the swamp. However, there are many different kinds of wetlands, including inland freshwater wetlands, coastal freshwater wetlands, and coastal saltwater wetlands.

IS THAT A FACT!

➤ About 6 percent of Earth's surface is wetlands.

➤ Wetland areas in the contiguous United States have shrunk from 81 million hectares to 38 million hectares and continue to shrink each year.

➤ Ninety percent of the wetlands in the San Francisco Bay area have disappeared. In their place are airports, houses, industrial parks, and landfills.

> **For background information about teaching strategies and issues, refer to the *Professional Reference for Teachers.***

CHAPTER 3

The Earth's Ecosystems

 Pre-Reading Questions

Students may not know the answers to these questions before reading the chapter, so accept any reasonable response.

Suggested Answers

1. There are many differences between a rain forest and a desert, such as water availability and average temperature.

2. Water in a lake is the endpoint of a river. The river may begin as snowmelt or may come from a spring, and it may grow from more and more tributaries joining it as it flows downstream to form a lake.

3. The open ocean contains few species, but some of them are quite large, such as the marine mammals. A swamp has many more species per unit area than does the open ocean.

CHAPTER 3

The Earth's Ecosystems

Sections

 **Pre-Reading Questions**

1. What are the main differences between a desert and a rain forest?

2. Where does the water in a lake come from?

3. Which has more species of plants and animals—the open ocean or a swamp? Why?

46

IN LIVING COLOR

A flurry of orange fish swim through the sun-dappled crevices of a tropical coral reef. All around them other life exists—sea fans, eels, anemones, and living corals. Could this scene exist anywhere else? It could . . . if the place was underwater in a warm climate close to the shore. In this chapter, you will learn how the nonliving environment affects organisms and how they are adapted to where they live.

START-UP Activity

A MINI-ECOSYSTEM

In this activity, you will build and observe a miniature ecosystem.

Procedure

1. Place a layer of **gravel** in the bottom of a **large widemouth jar** or **2 L bottle** with the top cut off. Add a layer of **soil.**

2. Add a variety of **small plants** that require similar growing conditions. Choose plants that will not grow too quickly.

3. Spray **water** inside the jar to moisten the soil.

4. Cover the jar, and place it in indirect light. Describe the appearance of your ecosystem in your ScienceLog.

5. Observe your mini-ecosystem every week. Spray it with water to keep the soil moist. Record all of your observations.

Analysis

6. List all of the nonliving factors in the ecosystem you have created.

7. How is your mini-ecosystem similar to a real ecosystem? How is it different?

47

START-UP Activity

A MINI-ECOSYSTEM

MATERIALS
FOR EACH GROUP: • gravel • large widemouthed jar or 2 L soda bottle • soil • plants • water

Safety Caution

You may want to have students wear disposable gloves while handling plants. Plants and plant parts should be kept away from the face and eyes because they may scratch or otherwise cause irritation. Have students wash their hands after handling plants, seeds, and soil.

Answers to START-UP Activity

6. Answers to this question should include most of the nonliving factors one would find in nature, such as water, temperature, light, and air.

7. Answers will vary. In general, each mini-ecosystem is similar to a real ecosystem in that it contains various abiotic and biotic factors that interact and affect each other. It will differ from a real ecosystem in that it is enclosed, it was artificially assembled, and it did not naturally evolve to its present state.

Land Ecosystems

Terms to Learn

abiotic	desert
biome	tundra
savanna	permafrost

What You'll Do

◆ Define *biome.*

◆ Describe three different forest biomes.

◆ Distinguish between temperate grasslands and savannas.

◆ Describe the importance of permafrost to the arctic tundra biome.

This section introduces the concept of a *biome* and describes several land biomes. Students learn about different types of forest and grassland as well as about deserts and tundra.

🔔 Bellringer

Write the following on the board:

cactus	tropical rain forest
tree frog	polar ice
pine tree	desert
polar bear	mountain

Ask students to match the plant or animal in the first column with the environment in the second column where it would most likely be found. They should record their answers in their ScienceLog. (Answer: cactus, desert; tree frog, tropical rain forest; pine tree, mountain; polar bear, polar ice)

① Motivate

ACTIVITY

Describing Ecosystems Pair up students and have them brainstorm about what distinguishes different land ecosystems—what makes a forest a forest, a desert a desert, and so on. Have the pairs write down their ideas and read them to the class. Write each idea on the board, and place marks next to an idea each time students suggest it. Tell students that they will learn how accurate this popularity contest was as they read this section.

Imagine that you are planning a camping trip. You go to a travel agency, where you find a virtual-reality machine that can let you experience different places before you go. You put on the virtual-reality gear, and suddenly you are transported. At first your eyes hurt from the bright sunlight. The wind that hits your face is very hot and very dry. As your eyes grow accustomed to the light, you see a large cactus to your right and some small, bushy plants in the distance. A startled jack rabbit runs across the dry, dusty ground. A lizard basks on a rock. Where are you?

You may not be able to pinpoint your exact location, but you probably realize that you are in a desert. That's because most deserts are hot and dry. These **abiotic,** or nonliving, factors influence the types of plants and animals that live in the area.

The Earth's Biomes

A desert is one of Earth's biomes. A **biome** is a geographic area characterized by certain types of plant and animal communities. A biome contains a number of smaller but related ecosystems. For example, a tropical rain forest is a biome that contains river ecosystems, treetop ecosystems, forest-floor ecosystems, and many others. A biome is not a specific place. For example, a desert biome does not refer to a particular desert. A desert biome refers to any and all desert ecosystems on Earth. The major biomes of Earth are shown in **Figure 1.**

Figure 1 *Rainfall and temperature are the main factors that determine what biome is found in a region. What kind of biome do you live in?*

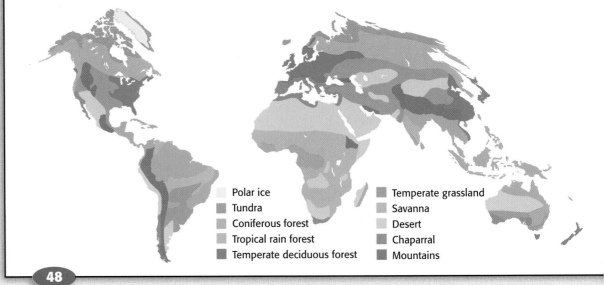

☐ Polar ice	■ Temperate grassland
■ Tundra	■ Savanna
■ Coniferous forest	■ Desert
■ Tropical rain forest	■ Chaparral
■ Temperate deciduous forest	■ Mountains

48

BRAIN FOOD

Have students examine the map of biomes in **Figure 1.** Ask, "Does every continent include a biome?" and "Where on the map is Antarctica?" Encourage students to speculate about why Antarctica does not include a biome. Ask them what kinds of plants and animals live in Antarctica. (Students may suggest seals, algae, bacteria, etc.)

Point out that most organisms that inhabit Antarctica are part of an aquatic biome rather than a land biome.

Forests

Forest biomes develop where there is enough rain and where the temperature is not too hot in the summer or too cold in the winter. There are three main types of forest biomes—temperate deciduous forests, coniferous forests, and tropical rain forests. The type of forest that develops depends on the area's temperature and rainfall.

Temperate Deciduous Forests In the autumn, have you seen leaves that change colors and fall from trees? If so, you have seen trees that are *deciduous*, which comes from a Latin word meaning "to fall off." By losing their leaves in the fall, deciduous trees are able to conserve water during the winter. **Figure 2** shows a temperate deciduous forest. Most of these forests contain several different species of trees. Temperate deciduous forests also support a variety of animals, such as bears and woodpeckers.

Temperate Deciduous Forest

Average Yearly Rainfall
75–125 cm (29.5–49 in.)

Average Temperatures
Summer: 28°C (82.4°F)
Winter: 6°C (42.8°F)

Figure 2 *In a temperate deciduous forest, mammals, birds, and reptiles thrive on the abundance of leaves, seeds, nuts, and insects.*

In forests, plant growth occurs in layers. The leafy tops of the trees reach high above the forest floor, where they receive full sunlight.

Beneath the tree layer, woody shrubs and bushes catch the light that filters through the trees.

Grasses, herbs, ferns, and mosses are scattered across the forest floor. Most of the flowering plants bloom, and produce seeds in early spring, before the trees grow new leaves.

49

COOPERATIVE LEARNING

Provide small groups of students with an almanac and graph paper. Ask groups to find and graph the average monthly rainfall and average monthly temperature for their area. Some students can measure the temperature and rainfall for 1 month. If this is not practical, encourage students to obtain this information from a local weather station or news broadcast. Have other students graph and present the data. You may also want to have students compare the climate of the region in which you live with that of the different biomes discussed in this lesson.

MATH and MORE

Have students compare the amount of rainfall in each of the forest biomes described. Then ask:

Which forest biome receives the most rain? (tropical rain forest)

How much more rain does this biome usually receive than the forest biome that receives the least amount of rain? (The coniferous forest receives a minimum of 35 cm per year. The tropical rain forest receives as much as 400 cm per year. The difference is 365 cm.)

 Math Skills Worksheet
"Subtraction Review"

 Teaching Transparency 73
"Coniferous Forest Biome"

Coniferous Forest

Average Yearly Rainfall
35–75 cm (14–29.5 in.)

Average Temperatures
Summer: 14°C (57.2°F)
Winter: −10°C (14°F)

Figure 3 *Many animals that live in a coniferous forest survive the harsh winters by hibernating or migrating to a warmer climate for the winter.*

A coniferous forest is home to many insects and to birds that eat them.

Herbivores that live in the coniferous forest include moose, deer, porcupines, and chipmunks.

These conifer leaves are adapted to conserve water.

Carnivores of the coniferous forest include foxes and lynxes.

50

Coniferous Forests Coniferous forests do not change very much from summer to winter. They are found in areas with long, cold winters. These forests consist mainly of *evergreen* trees, which are trees that don't lose their leaves and stay green all year. Most of these trees are *conifers,* which means that they produce seeds in cones. You have probably seen a pine cone. Pine trees are common conifers.

Most conifers can also be identified by their compact, needlelike leaves. These leaves, or needles, have a thick waxy coating that prevents them from drying out and being damaged during winter.

Figure 3 shows a coniferous forest and some of the animals that live there. Notice that not many large plants grow beneath the conifers, partly because very little light reaches the ground.

IS THAT A FACT!

The coniferous forest is also known as a boreal forest or a taiga. More of Earth's land can be categorized under this biome than under any other biome.

Tropical Rain Forests The tropical rain forest has more biological *diversity* than any other biome on the planet; that is, it contains more species than any other biome. As many as 100 species of trees may live in an area about one-fourth the size of a football field. Although some animals live on the ground, the treetops, or *canopy,* are the preferred living site. A huge variety of animals live in the canopy. If you counted the birds in the canopy of a rain forest, you would find up to 1,400 species! **Figure 4** shows some of the diversity of the tropical rain forest biome.

Most of the nutrients in a tropical rain forest biome are in the vegetation. The topsoil is actually very thin and poor in nutrients. Farmers who cut down the forest to grow crops must move their crops to freshly cleared land after about 2 years.

Tropical Rain Forest
Average Yearly Rainfall
Up to 400 cm (157.5 in.)
Average Temperatures
Daytime: 34°C (93°F)
Nighttime: 20°C (68°F)

Figure 4 A Tropical Rain Forest Biome

Trees of various heights form a continuous green roof, called the canopy, that may extend 60 m above the forest floor.

Woody vines climb the tree trunks to reach sunlight.

Little light reaches the ground. Low-growing plants of the rain forest don't require a lot of light.

51

USING THE FIGURE

Have students compare **Figure 2, Figure 3,** and **Figure 4** and locate each type of forest on the biome map in **Figure 1.** Ask students to focus on the different kinds of trees and animals in each forest picture. Encourage students to relate the different plants and animals to the differences in temperature and rainfall for each area. Sheltered English

REAL-WORLD CONNECTION

Biological diversity is an important natural resource. Provide the following example for students. About 40 percent of prescription drugs sold in the United States are derivatives of chemicals found in wild plants. Thousands more helpful plant compounds probably lie undiscovered in forests throughout the world. However, less than 1 percent of known plants have been tested for their medicinal effectiveness. Have students discuss the consequences of losing tropical rain forests given these considerations.

Teaching Transparency 74 "A Tropical Rain Forest Biome"

Math Skills Worksheet "Rain-Forest Math"

Multicultural CONNECTION

Scientist and Cornell University professor Eloy Rodriguez studies the chemical properties of plants to discover potentially useful compounds. One aspect of his research is learning about plants used by native cultures of tropical Africa, Asia, and Latin America. Using local knowledge may help Rodriguez find plant compounds with medicinal properties.

Temperate grasslands provide almost ideal growing conditions for grain crops. For this reason, few temperate grasslands remain today. In fact, these areas, such as the American Midwest and Ukraine, are sometimes called the breadbaskets of the world because their temperate grasslands became farmland for grain crops.

USING SCIENCE FICTION

Encourage students to read "The Greatest Asset," by Isaac Asimov. It can be found in the *Holt Anthology of Science Fiction*.

CONNECT TO
EARTH SCIENCE

Mountains can influence the climate of surrounding land, resulting in very different ecosystems existing in close proximity. Use the following Teaching Transparency to illustrate how a mountain can affect climate.

Teaching Transparency 177
"An Example of the Rain Shadow Effect"

LINK TO EARTH SCIENCE

---

internet connect

SciLINKS NSTA

TOPIC: Grasslands
GO TO: www.scilinks.org
*sci*LINKS **NUMBER:** HSTL485

Grasslands

Plains, steppes, savannas, prairies, pampas—these are names for regions where grasses are the major type of vegetation. Grasslands are found between forests and deserts. They exist on every continent. Most grasslands are flat or have gently rolling hills.

Temperate Grasslands Temperate grassland vegetation is mainly grasses mixed with a variety of flowering plants. There are few trees because fires prevent the growth of most slow-growing plants. The world's temperate grasslands support small, seed-eating mammals, such as prairie dogs and mice, and large herbivores, such as the bison of North America, shown in **Figure 5.**

Temperate Grassland
Average Yearly Rainfall
25–75 cm (10–29.5 in.)
Average Temperatures
Summer: 30°C (86°F)
Winter: 0°C (32°F)

Figure 5 *Bison roamed the temperate grasslands in great herds before they were hunted nearly to extinction.*

Savanna The **savanna** is a tropical grassland with scattered clumps of trees. During the dry season, the grasses die back, but the deep roots survive even through months of drought. During the wet season, the savanna may receive as much as 150 cm of rain. The savannas of Africa are inhabited by the most abundant and diverse groups of large herbivores in the world, like those shown in **Figure 6.** These include elephants, giraffes, zebras, gazelles, and wildebeests.

Savanna
Average Yearly Rainfall
150 cm (59 in.)
Average Temperatures
Dry season: 34°C (93°F)
Wet season: 16°C (61°F)

Figure 6 *Carnivores, such as lions and leopards, prey on herbivores, such as these zebras and wildebeests. Hyenas and vultures usually "clean up" after the carnivores.*

 BRAIN FOOD

During the dry season, savanna plants survive as roots. Ask students to discuss how herds of large herbivores might survive when the vegetation dries up.

(Students may suggest that the animals migrate to an area where food and water are more plentiful. Point out that migration is a behavioral adaptation for survival.)

Self-Check

Use the map in Figure 1 to compare the locations of deciduous and coniferous forests. Explain the differences in location between the two biomes. *(See page 168 to check your answers.)*

How do animals survive in the heat of the desert? Quite nicely, thank you! See how on page 132 of your LabBook.

3 Extend

 **PG 132**

"Life in the Desert"

Deserts

Deserts are hot, dry regions that support a variety of plants and animals. In a desert, most of the water that falls to the ground evaporates. Organisms have evolved in specialized ways to survive extreme temperatures with very little water. For example, plants grow far apart to reduce competition for the limited water supply. Some plants have shallow, widespread roots that absorb water quickly during a storm, while others may have very deep roots that reach ground water.

Animals also have adaptations for survival in the desert. Most are active only at night, when temperatures are cooler. Tortoises eat the flowers or leaves of plants and store the water under their shells for months. **Figure 7** shows how some desert plants and animals survive in the heat with little water.

Desert
Average Yearly Rainfall
Less than 25 cm (10 in.)
Average Temperatures
Summer: 38°C (100°F)
Winter: 7°C (45°F)

GROUP ACTIVITY

Divide the class into groups of four or five students. Then have students play "What Biome Am I?" Each student chooses a land biome. Then students take turns describing their biome to the other members of the group. Members of the group try to guess which biome the student is describing. The successful guesser takes the next turn. Allow students to play the game until everyone has had a chance to describe a biome.

USING THE FIGURE

Have students locate the chaparral biome in **Figure 1.** Tell them that the chaparral has a Mediterranean climate. Ask students to hypothesize about what characteristics the chaparral might have. (It is located on the coast and has a mild climate.)

Explain that many chaparral plants are adapted to survive drought and fire. The plants grow back after fires from small bits of surviving tissue.

Figure 7 *There are many well-adapted residents of the desert biome.*

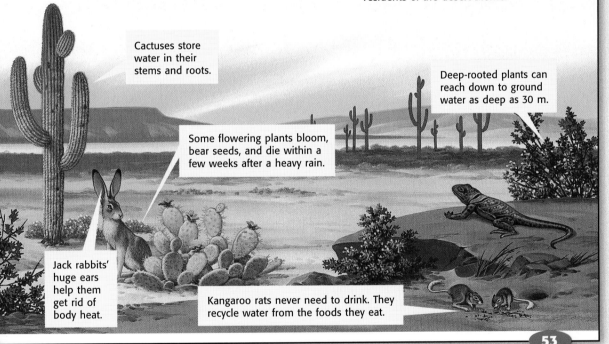

Cactuses store water in their stems and roots.

Deep-rooted plants can reach down to ground water as deep as 30 m.

Some flowering plants bloom, bear seeds, and die within a few weeks after a heavy rain.

Jack rabbits' huge ears help them get rid of body heat.

Kangaroo rats never need to drink. They recycle water from the foods they eat.

 53

Q: Which biome was Elvis's favorite?

A: Grassland

Answer to Self-Check

Deciduous forests tend to exist in mid-latitude, or temperate, regions, while coniferous forests tend to exist in colder, higher latitudes, closer to the poles.

Reinforcement Worksheet
"Know Your Biomes"

Ask students whether these statements are true or false. Have students correct any false statements they find.

1. Permafrost thaws only briefly in the summer. (false)
2. Tropical rain forests have more species than any other biome. (true)
3. Grasslands have very poor soil. (false)

ALTERNATIVE ASSESSMENT

Poster Project Have students choose one biome and make a poster to display facts and images of the biome. Posters should provide information about the uniqueness of the biome, indicate where the biome can be found, and compare the biome with other biomes.

Answer to MATHBREAK

Students should make a bar graph that includes the following data:

(The rainfall values below represent the top end of the average for each biome.)

Biome	Avg. yearly rainfall
Temp. deciduous forest	125 cm
Coniferous forest	75 cm
Tropical rain forest	400 cm
Temperate grassland	75 cm
Savanna	150 cm
Desert	25 cm
Tundra	50 cm

Tundra	
Average Yearly Rainfall	30–50 cm (12–20 in.)
Average Temperatures	
Summer: 12°C (53.6°F)	
Winter: −26°C (−14°F)	

Figure 8 *Caribou migrate to more plentiful grazing grounds during long, cold winters in the tundra.*

÷ 5 ÷ Ω ≤ ∞ +Ω √ 9 ∞ Σ 2

MATH BREAK

Rainfall

In 1 year, what is the difference in the rainfall amounts in a coniferous forest, a tropical rain forest, a desert, and a savanna? To compare, create a bar graph of the rainfall in each biome from the data given in this section.

54

Tundra

In the far north and on the tops of high mountains, the climate is so cold that no trees can grow. A biome called the **tundra** is found there.

Arctic Tundra The major feature of the arctic tundra is permafrost. During the short growing season, only the surface of the soil thaws. The soil below the surface, the **permafrost,** stays frozen all the time. Even though there is little rainfall, water is not in short supply. That's because the permafrost prevents the rain that does fall from draining, and the surface soil stays wet and soggy. Lakes and ponds are common.

The layer of unfrozen soil above the permafrost is too shallow for deep-rooted plants to survive. Grasses, sedges, rushes, and small woody shrubs are common. A layer of mosses and lichens grows beneath these plants on the surface of the ground. Tundra animals, like the one shown in **Figure 8,** include large mammals such as caribous, musk oxen, and wolves, as well as smaller animals, such as lemmings, shrews, and hares. Migratory birds are abundant in summer.

Alpine Tundra Another tundra biome is found above the tree line of very high mountains. These areas, called alpine tundra, receive a lot of sunlight and precipitation, mostly in the form of snow.

SECTION REVIEW

1. How is the climate of temperate grasslands different from that of savannas?
2. Describe three ways that plants and animals are adapted to the desert climate.
3. Where are most of the nutrients in a tropical rain forest?
4. **Applying Concepts** Could arctic tundra accurately be called a frozen desert? Why or why not?

▼ Answers to Section Review

1. Savannas are hotter than temperate grasslands and receive twice as much annual rainfall. Also rain falls only during the wet season in the savanna.
2. Answers will vary. Sample responses: being active only at night, developing structures that help them get rid of heat, and recycling consumed water.
3. Most of the biome's nutrients are contained in the vegetation.
4. Accept any logical, well-argued response. Sample answer: Yes, it would be accurate because very little rain falls and because liquid water is only available during the short growing season.

Terms to Learn

marine   zooplankton
phytoplankton   estuary

What You'll Do

- Distinguish between the different areas of the ocean.
- Explain the importance of plankton in marine ecosystems.
- Describe coral reefs and intertidal areas.

Marine Ecosystems

They cover almost three-quarters of Earth's surface and contain almost 97 percent of Earth's water supply. The largest animals on Earth inhabit them, along with billions of microscopic creatures, shown in **Figure 9.** Their habitats range from dark, cold, high-pressure depths to warm sandy beaches; from icy polar waters to rocky coastlines. They are oceans and seas. Wherever these salty waters are found, marine ecosystems are found. A **marine** ecosystem is one that is based on salty water. This abiotic factor has a strong influence on the ecosystems of oceans and seas.

Abiotic Factors Rule

Like terrestrial biomes, marine biomes are shaped by abiotic factors. These include temperature, the amount of sunlight penetrating the water, the distance from land, and the depth of the water. These abiotic factors are used to define certain areas of the ocean. As with terrestrial biomes, marine biomes occur all over Earth and can contain many ecosystems.

Sunny Waters Water absorbs light, so sunlight can penetrate only about 200 m below the ocean's surface, even in the clearest water. As you know, most producers use photosynthesis to make their own food. Because photosynthesis requires light, most producers are found only where light penetrates. The most abundant producers in the ocean are called **phytoplankton.** Phytoplankton are microscopic photosynthetic organisms that float near the surface of the water. Using the energy of sunlight, these organisms make their own food just as plants that live on land do. **Zooplankton** are the consumers that feed on the phytoplankton. They are small animals that, along with phytoplankton, form the base of the oceans' feeding relationships.

Figure 9 *Marine ecosystems support a broad diversity of life, from the humpback whale to microscopic phytoplankton.*

55

IS THAT A FACT!

The sea kelp known as *Macrocystis* can grow as much as 60 m in one growing season. This organism can grow lengthwise faster than any other organism known! It grows along the coast of California.

Directed Reading Worksheet Section 2

Marine Ecosystems

This section introduces different areas of the ocean as ecosystems. Students learn about the importance of abiotic factors in structuring oceanic life zones. The section also focuses on several unique marine ecosystems, including coral reefs and estuaries.

🔔 Bellringer

Write the following on the board or an overhead projector, and have students record their answers in their ScienceLog:

How much of the Earth's surface is covered by ocean?

$\frac{1}{4}$ $\frac{1}{2}$ $\frac{3}{4}$ $\frac{9}{10}$

(Answer: $\frac{3}{4}$)

What percentage of the Earth's water supply is found in the oceans?

37%   67%   77%   97%

(Answer: 97 percent)

1 Motivate

DISCUSSION

Oceans Students may not realize how connected their lives are to the oceans, especially if they live far from the coast. Ask students:

If all marine life were to die off, how would you be affected? How would others be affected?

Possible answers include: less food available (fish, crab, shrimp, shellfish, etc.), less oxygen in atmosphere (phytoplankton produce one-third to one-half of our oxygen), increased global warming (all that phytoplankton uses up huge amounts of carbon dioxide, the chief greenhouse gas), loss of jobs, and less enjoyment from ocean wildlife.

Multicultural CONNECTION

The Japanese and Koreans use brown seaweed to make *kombu* soup. Red algae is wrapped around rice to make a kind of sushi called *nori* rolls. Seaweed does provide iodine and other minerals, but it is used mostly for the taste and texture it adds to foods.

READING STRATEGY

Prediction Guide Before students read this page, ask them the following questions:

1. How do organisms in the dark depths of the ocean find food?
2. What kinds of organisms live on the ocean floor?
3. What is a thermal vent?
4. Which part of the ocean receives the most sunlight?

Have students evaluate their responses after they read about ocean biomes.

PG 133

Discovering Mini-Ecosystems

Wonderful Watery Biomes

Unique and beautiful biomes exist in every part of oceans and seas. These biomes are home to many unusually adapted organisms. The major ocean areas and some of the organisms that live in them are shown below in **Figure 10.**

A **The Intertidal Zone** The intertidal zone is the area where the ocean meets the land. This area is above water part of the day, when the tide is out, and is often battered by waves. Mud flats, rocky shores, and sandy beaches are all in the intertidal area.

B **The Neritic Zone** Moving seaward, the water becomes gradually deeper toward the edge of the continental shelf. Water in this area is generally less than 200 m deep and usually receives a lot of sunlight. Diverse and colorful coral reefs exist in the waters over the continental shelf, where the water is warm, clear, and sunny.

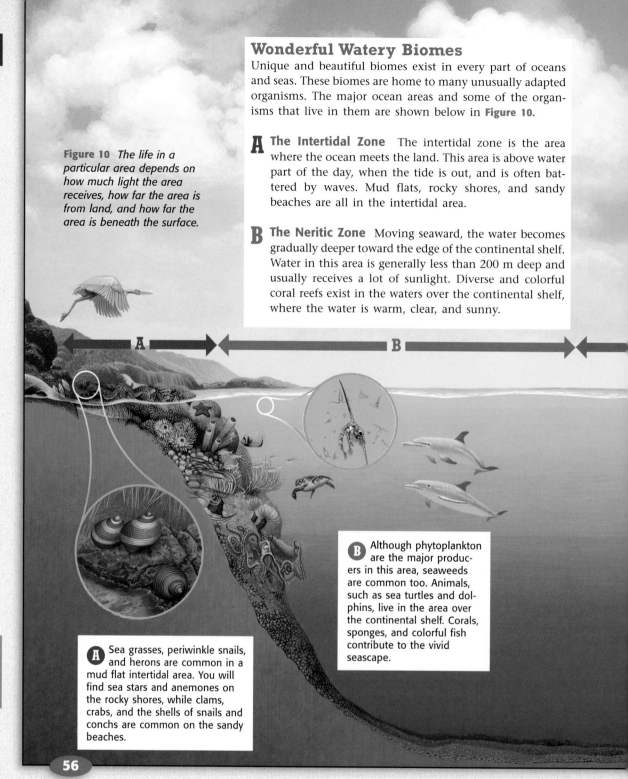

Figure 10 *The life in a particular area depends on how much light the area receives, how far the area is from land, and how far the area is beneath the surface.*

B Although phytoplankton are the major producers in this area, seaweeds are common too. Animals, such as sea turtles and dolphins, live in the area over the continental shelf. Corals, sponges, and colorful fish contribute to the vivid seascape.

A Sea grasses, periwinkle snails, and herons are common in a mud flat intertidal area. You will find sea stars and anemones on the rocky shores, while clams, crabs, and the shells of snails and conchs are common on the sandy beaches.

56

WEIRD SCIENCE

Food is scarce in the benthic zone. As a result, organic matter that filters down from the surface is often quickly consumed. Scavengers, including relatives of the freshwater shrimp, will voraciously attack a dead fish that falls to the deep ocean floor, reducing the carcass to bones in just a few hours!

C **The Oceanic Zone** Past the continental shelf, the sea floor drops sharply. This is the deep water of the open ocean. To a depth of about 200 m, phytoplankton are the producers. At greater depths, no light penetrates, so most organisms obtain energy by consuming organic material that falls from the surface.

D **The Benthic Zone** The benthic zone is the sea floor. It extends from the upper edge of the intertidal zone to the bottom of the deepest ocean waters. Organisms that live on the deep-sea floor obtain food mostly by consuming material that filters from above. Some bacteria are *chemosynthetic,* which means they use chemicals in the water near thermal vents to make food. A thermal vent is a place on the ocean floor where heat escapes through a crack in the Earth's crust.

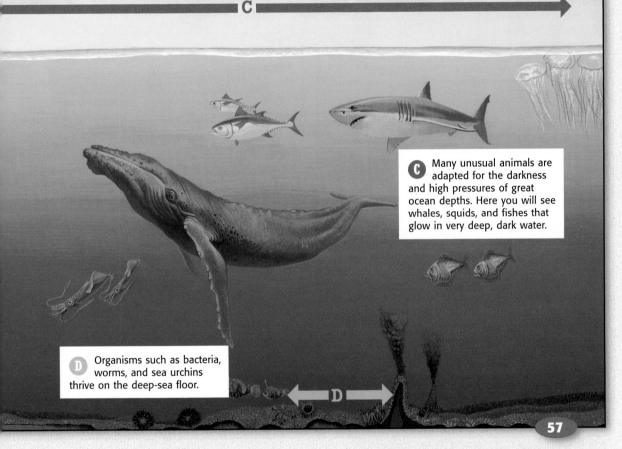

C Many unusual animals are adapted for the darkness and high pressures of great ocean depths. Here you will see whales, squids, and fishes that glow in very deep, dark water.

D Organisms such as bacteria, worms, and sea urchins thrive on the deep-sea floor.

57

IS THAT A FACT!

Deep-sea-vent ecosystems depend on neither sunlight nor photosynthesis for energy.

internetconnect

SCI*LINKS*
NSTA

TOPIC: Marine Ecosystems
GO TO: www.scilinks.org
*sci***LINKS NUMBER:** HSTL490

USING THE FIGURE

Help students understand the information in **Figure 10** by pointing out how it is organized. Make sure that students see the sequence of lettered captions above and below the water and link these letters to the letters indicating parts of the ocean. Also point out how the depth of the ocean changes as you go farther away from the shore. You may also want to help students interpret the magnified views of the ocean floor. Point out that these bubbles represent an up-close view of the area they extend from. Sheltered English

REAL-WORLD CONNECTION

A diving condition called the bends, or decompression sickness, can occur when a person dives very deeply, ascends to the surface too quickly, or stays under the water for a long period of time. The condition occurs when gas forms bubbles in body tissues. Divers suffering from decompression sickness need treatment in a pressure-decompression chamber.

CONNECT TO
PHYSICAL SCIENCE

Red, orange, and yellow wavelengths of light striking the surface of the water are absorbed first. Blue and green wavelengths can penetrate the water more deeply. Thus, producers capable of using blue and green wavelengths of light for photosynthesis can live at greater depths.

GROUP ACTIVITY

Figure 11 *A coral reef is one of the most biologically diverse biomes.*

 Have students research and draw a coral reef. Provide reference materials, a large piece of butcher paper, markers, and map pencils. Encourage pairs of students to choose an organism that is part of a coral reef, to research the organism, and to draw it. Then students should place their organism in its habitat on the butcher-paper coral reef. You may want to have one coral reef per class or have all classes contribute to one large coral reef. Sheltered English

GOING FURTHER

Writing Tell students that Texas has started a "Rigs for Reefs" program in which oil companies are asked to donate oil-drilling equipment that would otherwise be dragged to shore and sold for scrap metal. The old oil rigs and platforms are easily converted to artificial reefs. Have students pretend that they are a member of the board for this or a similar organization. Then have students write a persuasive letter to an oil company explaining why the company should participate in the program.

 Science Skills Worksheet "Being Flexible"

Interactive Explorations CD-ROM "Sea Sick"

A Closer Look

Marine environments provide most of the water for Earth's rainfall through evaporation and precipitation. Ocean temperatures and currents have major effects on world climates and wind patterns. Humans harvest enormous amounts of food from the oceans and dump enormous amounts of waste into them. Let's take a closer look at some of the special environments that thrive in the ocean.

Coral Reefs In some sunny tropical waters, the sea floor contains coral reefs. Corals live in a close relationship with single-celled algae. The algae produce organic nutrients through photosynthesis. This provides food for the coral. The coral provide a place in the sun for the algae to live. The foundation of the reef is formed from coral skeletons that have built up over thousands of years. Coral reefs, like the one in **Figure 11,** are home to many marine species, including a large variety of brightly colored fish and organisms such as sponges and sea urchins.

Figure 12 *The Sargasso Sea is a spawning place for eels and home to a rich diversity of organisms.*

The Sargasso Sea In the middle of the Atlantic Ocean is a large ecosystem with no land boundaries. It is called the Sargasso Sea. *Sargassum* is a type of algae usually found attached to rocks on the shores of North America, but it forms huge floating rafts in the Sargasso Sea. Animals adapted to this environment live among the algae. Most of the animals are the same color as the *Sargassum.* Some even look like it! Why do you think this is so? Can you find a fish in **Figure 12?**

✓ Self-Check

1. List three factors that characterize marine biomes.
2. Describe one way organisms obtain energy at great depths in the open ocean.

(See page 168 to check your answers.)

58

Answers to Self-Check

1. Answers include: The amount of sunlight penetrating the water, the distance from land, the depth of the water, the salinity of the water, and the water's temperature.

2. There are several possible answers. Some organisms are adapted for catching prey at great depths; some feed on dead plankton and larger organisms that filter down from above; and some, such as the bacteria around thermal vents, make food from chemicals in the water.

Polar Ice The Arctic Ocean and the open waters surrounding Antarctica make up a very unusual marine biome—one that includes ice!

The icy waters are rich in nutrients from the surrounding landmasses. These nutrients support large populations of plankton. The plankton in turn support a great diversity of fish, birds, and mammals, as shown in **Figure 13.**

Figure 13 *Sea lions and penguins are some of the animals found on the shores of Antarctica.*

Estuaries An area where fresh water from streams and rivers spills into the ocean is called an **estuary.** The fresh water constantly mixes with the salt water of the sea. The amount of salt in an estuary changes frequently. When the tide rises, the salt content of the water rises. When the tide recedes, the water becomes fresher. The fresh water that spills into an estuary is rich in nutrients that are carried by water running off the land. Because estuaries are so nutrient-rich, they support large numbers of plankton, which provide food for many larger animals.

Intertidal Areas Intertidal areas include mudflats, sandy beaches, and rocky shores. Mud flats are home to many worms and crabs and the shorebirds that feed on them. Sandy beaches are also home to worms, clams, crabs, and plankton that live among the sand grains.

On rocky shores, organisms either have tough holdfasts or are able to cement themselves to a rock to avoid being swept away by crashing waves. **Figure 14** shows some of these animals.

Figure 14 *Sea stars can wedge themselves under a rock to keep from being washed out to sea.*

SECTION REVIEW

1. Explain how a coral reef is both living and dead.

2. Why do estuaries support such an abundance of life?

3. **Analyzing Relationships** Explain how the amount of light an area receives determines the kinds of organisms that live in the open ocean.

internet connect

SC*i*LINKS
NSTA

TOPIC: Marine Ecosystems
GO TO: www.scilinks.org
***sci*LINKS NUMBER:** HSTL490

4 Close

Quiz

Ask students whether these statements are true or false. Have them rewrite any false statements so that they are true.

1. Although beautiful to look at, coral reefs are not very biologically diverse compared with other aquatic biomes. (false)

2. Ocean temperatures and currents have major effects on world climates. (true)

3. Phytoplankton are the most abundant consumers in the ocean. (false)

4. An intertidal area is one where the ocean meets the land. (true)

5. Marine ecosystems include those consisting of salt water or fresh water. (false)

ALTERNATIVE ASSESSMENT

Concept Mapping Have students create a concept map using the following terms:

marine biome, ocean, coral reef, polar ice, estuaries, intertidal areas, algae, phytoplankton, zooplankton

Remind students that they must supply words that show clear connections between the terms used in their maps.

▼ *Answers to Section Review*

1. The foundation of a coral reef is made up of coral skeletons that have built up over thousands of years. Living coral animals are found at the outer edges of a reef near the surface.

2. Nutrients that run off from land into fresh water are deposited with the sediments in estuaries, making them nutrient-rich areas. They are also areas where juveniles can

find shelter among the vegetation from rough water, tides, and predators.

3. Water absorbs light, so light does not travel into the deeper parts of the ocean. Therefore, the deeper an organism lives in the ocean, the less that organism is able to rely on the energy of sunlight. Phytoplankton, for example, must live in the photic zone.

Focus

Freshwater Ecosystems

This section introduces fresh-water ecosystems. Students learn about the characteristics of rivers, streams, and ponds. Students learn what happens to a pond as the seasons change. Students also learn how to distinguish between two types of wetlands.

Bellringer

Have students write an answer in their ScienceLog to the following question:

What are four different fresh-water ecosystems? (Answers may include the following: stream, river, lake, marsh, pond, swamp, bog, creek.)

1 Motivate

ACTIVITY

Writing Have students write a short description of a personal freshwater experience (something they did at a stream, lake, or wetland). If they have not had such an experience, ask them to use their imagination, or perhaps a field trip can be arranged. Have the students focus on sensations they remember from the experience as well as on what they and others were doing at the time.

Terms to Learn

tributary   wetland
littoral zone   marsh
open-water zone   swamp
deep-water zone

What You'll Do

◆ List the characteristics of rivers and streams.
◆ Describe the littoral zone of a pond.
◆ Distinguish between two types of wetlands.

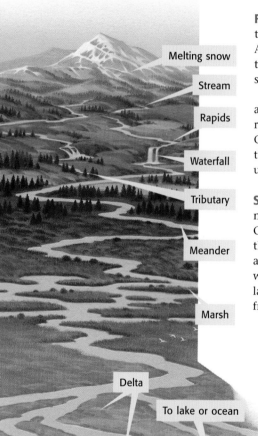

Melting snow
Stream
Rapids
Waterfall
Tributary
Meander
Marsh
Delta
To lake or ocean

60

Freshwater Ecosystems

A mountain brook bubbles over rocks down a mountainside. A mighty river thunders through a canyon. A small pond teems with life. A lake tosses boats during a heavy storm. A dense swamp echoes with the sounds of frogs and birds.

What do all of these places have in common? They are freshwater ecosystems. Like other ecosystems, freshwater ecosystems are characterized by abiotic factors, primarily the speed at which the water is moving.

Water on the Move

Brooks, streams, and rivers are ecosystems based on moving water. The water may begin flowing from melting ice or snow. Or it may come from a spring, where water flows up to the surface of the Earth. Each trickle or stream of water that joins a larger trickle or stream is a **tributary.**

Fast-Moving Water As more tributaries join a stream, the stream becomes larger and wider, forming a river. Aquatic plants line the edge of the river. Fishes live in the open waters. In the mud at the bottom, burrowers, such as freshwater clams and mussels, make their home.

Organisms that live in moving water require special adaptations to avoid being swept away with the current. Producers, such as algae and moss, cling to rocks. Consumers, such as insect larvae, live under rocks in the shallow water. Some consumers, such as tadpoles, use suction disks to hold themselves to rocks.

Slowing Down As a river grows wider and slower, it may *meander* back and forth across the landscape. Organic material and sediment may be deposited on the bottom, building *deltas.* Dragonflies, water striders, and other invertebrates live in and on slow-moving water. Eventually, the moving water empties into a lake or an ocean. **Figure 15** shows how a river can grow from melted snow.

Figure 15 *This figure shows the features of a typical river. Where is the water moving rapidly? Where is it moving slowly?*

Teaching Transparency 75
"River Features"

Directed Reading Worksheet Section 3

IS THAT A FACT!

The Nile River in Africa is the longest river in the world, at 6,670 km long. The Amazon River in South America comes in a close second, at 6,275 km long.

Still Waters

Ponds and lakes have different ecosystems than streams and rivers have. Lake Superior, the largest lake in the world, has more in common with a small beaver pond than with a river. **Figure 16** shows a cross section of a typical lake. In looking at this illustration, you will notice that the lake has been divided into three zones. As you read on, you will learn about these zones and the ecosystems they contain.

Where Water Meets Land Look at Figure 16 again, and locate the **littoral zone.** It is the zone closest to the edge of the land. This zone has many inhabitants. Plants that grow in the water closest to the shore include cattails and rushes. Farther from the shore are floating leaf plants, such as water lilies. Still farther out are submerged pond weeds that grow beneath the surface of the water.

The plants of the littoral zone provide a home for small animals, such as snails, small arthropods, and insect larvae. Clams, worms, and other organisms burrow in the mud. Frogs, salamanders, water turtles, various kinds of fishes, and water snakes also live in this area.

Life at the Top Look again at Figure 16. This time locate the **open-water zone.** This zone extends from the littoral zone across the top of the water. The open-water zone only goes as deep as light can reach. This is the habitat of bass, blue gills, lake trout, and other fish. Phytoplankton are the most abundant photosynthetic organisms in the open-water zone of a lake.

Life at the Bottom Now look at Figure 16 and find the **deep-water zone.** This zone is below the open-water zone, where no light reaches. Catfish, carp, worms, insect larvae, crustaceans, fungi, and bacteria live here. These organisms feed on dead organic material that falls down from above.

QuickLab
Pond Food Connections

1. On **index cards**, write the names of the animals and plants that live in a typical freshwater pond or small lake. Write one type of organism on each card.
2. Use **yarn** or **string** to connect each organism to its food sources.
3. In your ScienceLog, describe the food relationships in the pond.

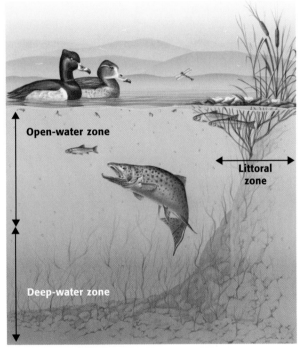

Open-water zone

Littoral zone

Deep-water zone

Figure 16 *Freshwater ecosystems are characterized by abiotic factors that determine which organisms live there.*

61

CONNECT TO PHYSICAL SCIENCE

Unlike most substances, solid water is less dense than liquid water. This is why ice floats. This property of water allows organisms to live beneath the frozen surface of the water. Ask students to consider what might happen if liquid water was less dense than frozen water. (Ice would sink, and entire bodies of water would freeze in winter, killing the organisms inhabiting the water.)

This should spark a lot of discussion in the classroom. Students should realize that a year-long expedition will span all four seasons. They should list gear that will help them survive the weather changes over the period. They should mention hunting and fishing as a means of providing food for themselves, and they should list the equipment necessary for those activities as items that they would bring along. They should also mention ways of sampling both land plants and aquatic plants and the equipment they would need to do so. Some problems they may encounter would probably be weather related.

MAKING MODELS

Have students design and construct a three-dimensional cut-away model of a lake. They should use whatever materials they wish to portray the three zones (littoral, open water, and deep water) and a sample of the plants and animals found in these zones.
Sheltered English

 internet**connect**

SCI**LINKS**
NSTA

TOPIC: Freshwater Ecosystems
GO TO: www.scilinks.org
*sci***LINKS NUMBER:** HSTL495

A Trip to Lake Superior
Suppose you are a life scientist who specializes in the plants that live in and near Lake Superior. You are preparing for a yearlong expedition to Thunder Bay, on the Canadian shore of Lake Superior.

You will stay "in the wild." Based on what you have learned about ecosystems, answer the following questions: How will you live while you are there? What will you bring along? What problems will you encounter? How will you overcome them?

Activity

While exploring in a wetland, you have discovered a new organism. In your ScienceLog, draw the organism. Describe what it looks like and how it is adapted to its environment. Trade with a partner. Is your partner's organism believable?

TRY at HOME

Wetlands

A **wetland** is an area of land where the water level is near or above the surface of the ground for most of the year. Wetlands support a variety of plant and animal life. They also play an important role in flood control. During heavy rains or spring snow melt, wetlands soak up large amounts of water. The water in wetlands also seeps into the ground, replenishing underground water supplies.

Marshes A **marsh** is a treeless wetland ecosystem where plants such as cattails and rushes grow. A freshwater marsh is shown in **Figure 17.** Freshwater marshes are found in shallow waters along the shores of lakes, ponds, rivers, and streams. The plants in a marsh vary depending on the depth of the water and the location of the marsh. Grasses, reeds, bulrushes, and wild rice are common marsh plants. Muskrats, turtles, frogs, and red-wing blackbirds can be found living in marshes.

Figure 17 *Turtles find a lot of places to escape from predators in a freshwater marsh. Many species raise their young in these protected areas.*

IS THAT A FACT!

Fish living in streams with strong currents face a problem: how do they prevent themselves from being washed downstream? Some have developed suckers that hold them in place. Some stay in relatively calm waters behind rocks, and at least two—a kind of catfish in the Andes and a kind of loach in Borneo—have developed huge lips that they use to clamp on to river debris to hold their position.

Swamps A **swamp** is a wetland ecosystem where trees and vines grow. Swamps occur in low-lying areas and beside slow-moving rivers. Most swamps are flooded only part of the year, depending on the rainfall. Trees may include willows, bald cypresses, water tupelos, oaks, and elms. Vines such as poison ivy grow up trees, and Spanish moss hangs from the branches. Water lilies and other lake plants may grow in open-water areas. Swamps, like the one in **Figure 18,** provide a home for a variety of fish, snakes, and birds.

Figure 18 *The bases of the trunks of these trees are adapted to give the tree more support in the wet, soft sediment under the water in this swamp.*

From Lake to Forest

How can a lake or pond, like the one in **Figure 19,** disappear? Water entering a standing body of water usually carries nutrients and sediment along with it. These materials then settle to the bottom. Dead leaves from overhanging trees and decaying plant and animal life also settle to the bottom. Gradually, the pond or lake fills in. Plants grow in the newly filled areas, closer and closer toward the center. With time, the standing body of water becomes a marsh. Eventually, the marsh turns into a forest.

Figure 19 *Eventually decaying organic matter, along with sediment in the runoff from land, will fill in this pond.*

SECTION REVIEW

1. Describe some adaptations of organisms that live in moving water.

2. Compare the littoral zone with the open-water zone of a pond.

3. How is a swamp different from a marsh?

4. **Analyzing Concepts** The center of a pond is 10 m deep. Near the shore it is 0–1 m deep. Describe the types of organisms that might live in each zone.

internetconnect

SC*i*LINKS.
NSTA

TOPIC: Freshwater Ecosystems
GO TO: www.scilinks.org
*sci*LINKS NUMBER: HSTL495

63

Quiz

Ask students whether these statements are true or false.

1. Freshwater ecosystems can be grouped as those that are still, those that are flowing, and those that contain salt. (false)

2. Catfish, carp, and other scavengers are likely to be found in the deep-water zone of a lake. (true)

3. Organisms living in fast-moving water do not usually exhibit any special adaptations to their environment. (false)

ALTERNATIVE ASSESSMENT

Writing Have students write a travel brochure for one of the ecosystems in this chapter. Brochures should provide information about climate, recreation, wildlife, and conservation efforts for the area. Encourage students to be informative as well as creative.
Sheltered English

Critical Thinking Worksheet
"Risky Development?"

▼ **Answers to Section Review**

1. Organisms that live in moving waters tend to use fixed objects, such as rocks, as habitats and/or stabilizers.

2. The littoral zone of a pond is shallower than the open water and contains more nutrients and vegetation. There is more

cover and protection from predators for juveniles in the littoral zone.

3. A marsh is a wetland without trees, and a swamp is a wetland with trees.

4. Student answers should reflect the information found on pages 61–63.

Skill Builder Lab

Too Much of a Good Thing?
Teacher's Notes

Time Required

One 45-minute class period and one 10-minute observation time every 3 days for 3 weeks

Lab Ratings

EASY			HARD

TEACHER PREP 🧪🧪

STUDENT SET-UP 🧪🧪

CONCEPT LEVEL 🧪🧪🧪

CLEAN UP 🧪🧪

MATERIALS

The materials listed on the student page are enough for 1–2 students. This lab is a good opportunity to recycle glass jars or clear plastic 2 L soda bottles. Any container that is transparent and will hold at least 1 L of water will do.

Safety Caution

Remind students to review all safety cautions and icons before beginning this lab activity.

Preparation Notes

A review of the causes of eutrophication might be helpful before beginning this lab.

Jason Marsh
Montevideo High and
Country School
Montevideo, Minnesota

Too Much of a Good Thing?

Plants require nutrients, such as phosphates and nitrates. Phosphates are often found in detergents. Nitrates are often found in animal wastes and fertilizers. When large amounts of these nutrients enter rivers and lakes, algae and plant life grow rapidly and then die off. Microorganisms that decompose the dead matter use up oxygen in the water, killing fish and other animals. In this activity, you will observe the effect of fertilizers on organisms that live in pond water.

MATERIALS

• wax pencil
• 1 qt (or 1 L) jars (3)
• 2.25 L of distilled water
• fertilizer
• graduated cylinder
• stirring rod
• 300 mL of pond water containing living organisms
• eyedropper
• microscope
• microscope slides with coverslips
• plastic wrap
• protective gloves

Procedure

1. Use a wax pencil to label one jar "Control," the second jar "Fertilizer," and the third jar "Excess Fertilizer."

2. Pour 750 mL of distilled water in each of the jars. Read the label on the fertilizer container to determine the recommended amount of fertilizer. To the Fertilizer jar, add the amount of fertilizer recommended for 750 mL of water. To the Excess Fertilizer jar, add 10 times the recommended amount. Stir the contents of each jar to dissolve the fertilizer.

3. Obtain a sample of pond water. Stir it gently but thoroughly to make sure that the organisms in it are evenly distributed. Pour 100 mL of pond water into each of the three jars.

4. Observe a drop of pond water from each jar under the microscope. Draw at least four of the organisms. Determine whether the organisms you see are algae, which are usually green, or consumers, which are usually able to move. Describe the number and type of organisms in the pond water.

Common Pond-Water Organisms

| *Volvox* (producer) | *Spirogyra* (producer) | *Daphnia* (consumer) | *Vorticella* (consumer) |

64

5 Cover each jar loosely with plastic wrap. Place the jars near a sunny window, but do not place them in direct sunlight.

6 Based on your understanding of how ponds and lakes eventually fill up to become dry land, make a prediction about how the pond organisms will grow in each of the three jars.

7 Make three data tables in your ScienceLog. Be sure to allow enough space to record your observations. Title one table "Control," as shown below. Title another table "Fertilizer," and title the third table "Excess Fertilizer."

8 Observe the jars when you first set them up and at least once every three days for the next three weeks. Note the color, odor, and any visible presence of organisms. Record your observations.

9 When organisms begin to be visible in the jars, use an eyedropper to remove a sample from each jar, and observe the sample under the microscope. How have the number and type of organisms changed since you first looked at the pond water? Record your observations.

10 At the end of the three-week period, remove a sample from each jar and observe each sample under the microscope. Draw at least four of the most abundant organisms, and describe how the number and type of organisms have changed since your last microscopic observation.

Analysis

11 After three weeks, which jar has the most abundant growth of algae? What may have caused this growth?

12 Did you observe any effects on organisms (other than the algae) in the jar with the most abundant algal growth? Explain your answer.

13 Did your observations match your prediction? Explain your answer.

14 How might the rapid filling of natural ponds and lakes be prevented or slowed?

Control			
Date	Color	Odor	Other observations

DO NOT WRITE IN BOOK

Answers

11. Answers will vary. Occasionally, the jar with excess fertilizer will grow beyond the carrying capacity of algae or other organisms and will crash.

12. Animal organisms occasionally die from the toxic effects of the fertilizer. When algae growth is overwhelming, the depletion of oxygen in the water will have a detrimental effect on animal organisms.

13. Answers will vary. Students should defend their prediction.

14. Answers will vary, but most students will say that we need to use less fertilizer. Eutrophication is continuously occurring and is a natural process. Fertilizers artificially hasten the process. Have students discuss how the global food supply might be affected if no fertilizers were used to grow crops. Then ask them to brainstorm to come up with some solutions.

 Datasheets for LabBook

65

Chapter Highlights

SECTION 1

abiotic describes nonliving factors in the environment

biome a large region characterized by a specific type of climate and certain types of plant and animal communities

savanna a tropical grassland biome with scattered clumps of trees

desert a hot, dry biome inhabited by organisms adapted to survive high daytime temperatures and long periods without rain

tundra a far-northern biome characterized by long, cold winters, permafrost, and few trees

permafrost the permanently frozen ground just below the surface of the soil in the arctic tundra

SECTION 2

marine an ecosystem based on salty water

phytoplankton microscopic photosynthetic organisms that float near the surface of the ocean

zooplankton very small animals that, along with the phytoplankton they consume, form the base of the oceans' food web

estuary an area where fresh water from streams and rivers spills into the ocean

Chapter Highlights

SECTION 1

Vocabulary

abiotic *(p. 48)*
biome *(p. 48)*
savanna *(p. 52)*
desert *(p. 53)*
tundra *(p. 54)*
permafrost *(p. 54)*

Section Notes

- Rainfall and temperature are the main factors that determine what kind of biome is found in a region.

- The three main forest biomes are the temperate deciduous forest and the coniferous forest, which experience warm summers and cold winters, and the tropical rain forest, where temperatures stay warm.

- Grasslands receive more rain than deserts and receive less rain than forests. Temperate grasslands have hot summers and cold winters. Savannas have wet and dry seasons.

- Deserts receive less than 25 cm of rain a year. Plants and animals competing for the limited water supply have developed special adaptations for survival.

- The tundra biome is found mainly in the Arctic region. Arctic tundra is characterized by permafrost.

Labs

Life in the Desert *(p. 132)*

SECTION 2

Vocabulary

marine *(p. 55)*
phytoplankton *(p. 55)*
zooplankton *(p. 55)*
estuary *(p. 59)*

Section Notes

- The kinds of marine organisms that inhabit an area vary depending on the water depth, the temperature, the amount of light, and the distance from shore.

- The intertidal area is the area where sea and land meet.

- The sea floor is home to biomes as different as coral reefs and thermal vents.

- The open ocean includes unique biomes, including the Sargasso Sea and the cold water oceans around the poles.

☑ Skills Check

Math Concepts

RAINFALL Using a meterstick, measure 400 cm on the floor of your classroom. This distance represents the depth of rainfall a rain forest receives per year. Next measure 25 cm. This measurement represents the amount of rainfall a desert receives per year. Compare these two quantities. Express your comparison as a ratio.

$$\frac{25}{400} = \frac{1}{16}$$

In 1 year, a desert receives $\frac{1}{16}$ the rainfall that a rain forest receives.

Visual Understanding

RAIN FOREST Look at Figure 4, on page 51. There are three layers of a rain forest—the upper story, the middle story, and the ground story. The upper story is the canopy, where most rain forest species live and where there is the most sunlight. The middle story is under the canopy and above the ground. The ground story is dark in most parts of the forest. Most plants in the rain forest grow very tall to compete for light in the canopy. Growth of plants on the ground story is not very dense due to the lack of available light.

Lab and Activity Highlights

Too Much of a Good Thing? PG 64

Life in the Desert PG 132

Discovering Mini-Ecosystems PG 133

Datasheets for LabBook
(blackline masters for these labs)

SECTION 2

- An estuary is a region where fresh water from rivers spills into the ocean and the fresh and salt water mix with the rising and falling of the tides.

Labs

Discovering Mini-Ecosystems
(p. 133)

SECTION 3

Vocabulary

    tributary *(p. 60)*

    littoral zone *(p. 61)*

    open-water zone *(p. 61)*

    deep-water zone *(p. 61)*

    wetland *(p. 62)*

    marsh *(p. 62)*

    swamp *(p. 63)*

Section Notes

- Freshwater ecosystems are classified according to whether they have running water or standing water. Brooks, rivers, and streams contain running water. Lakes and ponds contain standing water.

- As tributaries join a stream between its source and the ocean, the volume of water in the stream increases, the nutrient content increases, and the speed decreases.

- The types of organisms found in a stream or river are determined mainly by how quickly the current is moving.

- The littoral zone of a lake is inhabited by floating plants. These plants provide a home for a rich diversity of animal life.

- Wetlands include marshes, which are treeless, and swamps, where trees and vines grow.

VOCABULARY DEFINITIONS, *continued*

SECTION 3

tributary a small stream or river that flows into a larger one

littoral zone the zone of a lake or pond closest to the edge of the land

open-water zone the zone of a lake or pond that extends from the littoral zone out across the top of the water and that is only as deep as light can reach through the water

deep-water zone the zone of a lake or pond below the open-water zone, where no light reaches

wetland an area of land where the water level is near or above the surface of the ground for most of the year

marsh a treeless wetland ecosystem where plants such as cattails and rushes grow

swamp a wetland ecosystem where trees and vines grow

 Vocabulary Review Worksheet

Blackline masters of these Chapter Highlights can be found in the **Study Guide.**

---

internet**connect**

GO TO: go.hrw.com

Visit the **HRW** Web site for a variety of learning tools related to this chapter. Just type in the keyword:

KEYWORD: HSTECO

*SCI*LINKS™
N S T A

GO TO: www.scilinks.org

Visit the **National Science Teachers Association** on-line Web site for Internet resources related to this chapter. Just type in the *sci*LINKS number for more information about the topic:

TOPIC: Forests   **sci*LINKS NUMBER:** HSTL480
TOPIC: Grasslands   **sci*LINKS NUMBER:** HSTL485
TOPIC: Marine Ecosystems **sci*LINKS NUMBER:** HSTL490
TOPIC: Freshwater Ecosystems **sci*LINKS NUMBER:** HSTL495

67

Lab and Activity Highlights

LabBank

 EcoLabs & Field Activities, Biome Adventure Travel

Long-Term Projects & Research Ideas, Tropical Medicine

Interactive Explorations CD-ROM

 CD 2, Exploration 2, "Sea Sick"

Chapter Review
Answers

USING VOCABULARY

1. continental shelf
2. Zooplankton
3. marsh
4. Deciduous trees
5. tundra
6. biome

UNDERSTANDING CONCEPTS

Multiple Choice

7. a
8. c
9. b
10. d
11. d

Short Answer

12. Tributaries flow into it, increasing the volume of water. The salt and nutrient content of the water increases, as does the cloudiness. The speed of the water decreases.

13. Answers may vary; possible answer: being active only at night, and storing water in body tissue

14. No, wetlands are not *necessarily* always wet. However, wetlands have high water tables most of the year, so wetland soils are wet much of the time.

15. The salt content in the water of an estuary changes constantly as the tides rise and fall, moving water in and out of the estuary. The most important factor in the salinity of an estuary, however, is freshwater runoff from the land. After heavy rainfall, the salinity of the water in an estuary close to land will be very low because the salts will be diluted with additional fresh water.

Chapter Review

To complete the following sentences, choose the correct term from each pair of terms listed below:

1. At the edge of the __?__, the open ocean begins. *(continental shelf* or *Sargasso Sea)*

2. __?__ are tiny consumers that live in water. *(Phytoplankton* or *Zooplankton)*

3. A __?__ is a treeless wetland. *(swamp* or *marsh)*

4. __?__ lose their leaves in order to conserve water. *(Deciduous trees* or *Conifers)*

5. The major feature of the __?__ biome is permafrost. *(desert* or *tundra)*

6. Each major type of plant community and its associated animal communities make up a(n)__?__. *(estuary* or *biome)*

Multiple Choice

7. The most numerous organisms in the oceans are the
   a. plankton.
   b. *Sargassum.*
   c. coral animals.
   d. marine mammals.

8. Marine ecosystems at the poles are unusual because
   a. animals spend time both in and out of the water.
   b. plankton are rare.
   c. they contain ice.
   d. the salt content of the water is very high.

9. The major factor that determines the types of organisms that live in a stream or river is
   a. the water temperature.
   b. the speed of the current.
   c. the depth of the water.
   d. the width of the stream or river.

10. Marine ecosystems
    a. contain the largest animals in the world.
    b. exist in all ocean zones.
    c. include environments where organisms survive without light.
    d. All of the above

11. Two major factors that determine what kind of a biome is found in a region are
    a. the amount of rainfall and the temperature.
    b. the depth of water and the distance from land.
    c. the wave action and the salt content of the water.
    d. All of the above

Short Answer

12. Describe how a stream changes as it moves from its source toward the ocean.

13. Describe two adaptations of animals to the desert environment.

14. Are wetlands always wet? Explain.

15. Explain how the salt content in an estuary changes constantly.

68

Concept Mapping

16. Use the following terms to create a concept map: tropical rain forest, deep-rooted plants, coral reef, canopy, biomes, permafrost, desert, continental shelf, tundra, ecosystems.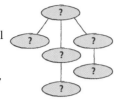

CRITICAL THINKING AND PROBLEM SOLVING

Write one or two sentences to answer the following questions:

17. While excavating a region now covered by grasslands, paleontologists discover the fossil remains of ancient fish and shellfish. What might they conclude?

18. In order to build a new shopping center, developers fill in a wetland. Afterward, flooding becomes a problem in this area. How can this be explained?

19. Explain why most desert flowering plants bloom, bear seeds, and die within a few weeks, while some tropical flowering plants remain in bloom for a much longer time.

MATH IN SCIENCE

20. What is the average difference in rainfall between a temperate deciduous forest and a coniferous forest?

21. An area of Brazilian rain forest received 347 cm of rain in one year. Using the following formula, calculate this amount of rainfall in inches.

0.394 (the number of inches in a centimeter)

× 347 cm

____?____ in.

INTERPRETING GRAPHICS

The graphs below show the monthly temperatures and rainfall in a region during 1 year.

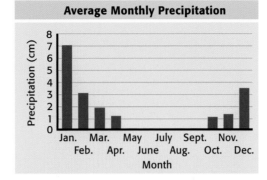
Average Monthly Precipitation

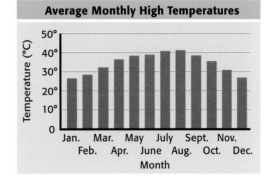
Average Monthly High Temperatures

22. What kind of biome is probably found in the region represented by these graphs?

23. Would you expect to find bulrushes in the region represented by these graphs? Why or why not?

 **Reading Check-up** Take a minute to review your answers to the Pre-Reading Questions found at the bottom of page 46. Have your answers changed? If necessary, revise your answers based on what you have learned since you began this chapter.

Concept Mapping

16. An answer to this exercise can be found at the front of this book.

CRITICAL THINKING AND PROBLEM SOLVING

17. At one time, this region was covered by water.

18. The wetland had soaked up large amounts of water. Because the water can no longer drain into the wetland, it now runs off the land, causing the flooding.

19. Rainfall is sparse in the desert. When rain does fall, flowering plants of the desert must bloom and bear seeds quickly, while there is still water in the ground.

MATH IN SCIENCE

20. about 45 cm
21. about 137 in.

INTERPRETING GRAPHICS

22. desert
23. No; The graph represents a typical hot, dry desert. Bulrushes grow in wetlands.

 Concept Mapping Transparency 20

 Blackline masters of this Chapter Review can be found in the **Study Guide**.

ACROSS THE SCIENCES

LIFE SCIENCE • CHEMISTRY

Ocean Vents

Background

The first black smoker was found off the Galápagos Islands in 1977, and since then at least four have been found on the bottom of the Pacific Ocean floor 322 km (200 mi) off the coast of Canada. The heat generated by one of them was so intense that the surface of the ocean water was nearly boiling when the chimney was lifted out.

It seems almost impossible that creatures on our planet could survive under such conditions, but one of the species of tube-worms can withstand temperatures up to 80°C (176°F)!

Other recently discovered creatures on the bottom of the ocean thrive on frozen chunks of methane in the same way that these creatures thrive on sulfur. These discoveries of chemosynthetic life have been called a "biological revolution" by one scientist and have sparked debate about the possibilities for life on other planets.

▲ *"They're very slim, fuzzy, flattened-out worms. Really hairy,"* says scientist *Bob Feldman about tubeworms.*

Picture the extreme depths of the ocean. There is no light at all, and it is very cold. But in the cracks between the plates on the bottom of the ocean floor, sea water trickles deep into the Earth. On the way back up from these cracks, the heated water collects metals, sulfuric gases, and enough heat to raise the temperature of the chilly ocean to 360°C. That is hot enough to melt lead! This heated sea water blasts up into the ocean through volcanic vents. And when this hot and toxic brew collides with icy ocean waters, the metals and sulfuric gases *precipitate,* that is, settle out of the heated ocean water as solids.

These solids form tubes, called black smokers, that extend up through the ocean floor. To humans, this dark, cold, and toxic environment would be deadly. But to a community of 300 species, including certain bacteria, clams, mussels, and tube worms, it is home. For these species, black smokers make life possible.

Life Without Photosynthesis

For a long time, scientists believed that energy from sunlight was the basis for the Earth's food chains and for life itself. But in the last 15 years, researchers have discovered ecosystems that challenge this belief. We now know of organisms around black smokers that can live without sunlight. One type of bacteria uses toxic gases from a black smoker in the same way that plants use sunlight. In a process called *chemosynthesis,* these bacteria convert sulfur into energy.

These bacteria are producers, and the mussels and clams are the consumers in this deep-sea food web. The bacteria use the mussels and clams as a sturdy place to live. The mussels and clams, in turn, feed off the bacteria. This kind of relationship between organisms is called *symbiosis.* The closer to the vent the clams and mussels are, the more likely the bacteria are to grow. Because of this, the mussels and clams frequently move to find good spots near the black smokers.

What Do You Think?

▶ Conditions near black smokers are similar to conditions on other planets. Do some research on these extreme environments, both on Earth and elsewhere. Then discuss with your classmates where and how you think life on Earth may have started.

70

Answer to What Do You Think?

Answers will vary. Accept any response that is well researched, thoughtful, and well articulated.

CAREERS

ECOLOGIST

Most winters **Alfonso Alonso-Mejía** climbs up to the few remote sites in central Mexico where about 150 million monarch butterflies spend the winter. He is researching the monarchs because he wants to help preserve their habitat.

Monarch butterflies are famous for their long-distance migration. Those that eventually find their way to Mexico come from as far away as the northeastern United States and southern Canada. Some of them travel 3,200 km before reaching central Mexico.

Human Threats to Habitats

Unfortunately, the monarchs' habitat is increasingly threatened by logging and other human activities. Only nine of the monarchs' wintering sites remain. Five of the sites are set aside as sanctuaries for the butterflies, but even those are endangered by people who cut down fir trees for firewood or for commercial purposes.

Research to the Rescue

Alonso-Mejía's work is helping Mexican conservationists better understand and protect monarch butterflies. Especially important is his discovery that monarchs depend on bushlike vegetation that grows beneath the fir trees, called understory vegetation.

Alonso-Mejía's research showed that when the temperature dips below freezing, as it often does at the high-altitude sites where the monarchs winter, some monarchs depend on understory vegetation for survival. This is because low temperatures ($-1°C$ to $4°C$) limit the monarchs' movement—the butterflies are not even able to crawl. At extremely cold temperatures ($-7°C$ to $-1°C$), monarchs resting on the forest floor are in danger of freezing to death. But where there is understory vegetation, the monarchs can slowly climb the vegetation until they are at least 10 cm above the ground. This tiny difference in elevation can provide a microclimate that is warm enough to ensure the monarchs' survival.

The importance of understory vegetation was not known before Alonso-Mejía did his research. Now, thanks to his work, Mexican conservationists will better protect the understory vegetation.

Get Involved!

▶ If you are interested in a nationwide tagging program to help scientists learn more about the monarchs' migration route, write to Monarch Watch, Department of Entomology, 7005 Howorth Hall, University of Kansas, Lawrence, Kansas 66045.

Background

Why do animals migrate? Animals migrate, or move from one area to another area, to take advantage of different environmental conditions in different locations. Monarch butterflies, like many birds, migrate north in the summers, when the food supply and climate are favorable, and south in the winters, when the climate in the north is unfavorable.

Migrating animals often inhabit distinctly different biomes, as in the case of the monarch butterfly. Given their unique ecology, migrating animals can serve as indicators of the health of different ecosystems. Indeed, monarch butterflies face challenges in both their summer and winter homes: loss of habitat in Mexico and loss of their primary food source, milkweed, in Canada.

71

Discussion

Lead your students in a discussion about what Alfonso's research suggests about the organisms that inhabit Earth's ecosystems. Even common, broad-ranging organisms such as the monarch butterfly often have specific requirements for food and shelter. Monarch butterflies do not just need forests to winter in; they also need cool, humid forests covered with specific types of upperstory and understory vegetation. The monarch's tenuous situation illustrates the importance of both biotic (vegetation) and abiotic (temperature) factors for an organism's survival.

Chapter Organizer

CHAPTER ORGANIZATION	TIME MINUTES	OBJECTIVES	LABS, INVESTIGATIONS, AND DEMONSTRATIONS
Chapter Opener **pp. 72–73**	45	National Standards: ST 1, 2, SPSP 5, HNS 1, 2, LS 3a, 5c	**Start-Up Activity,** Recycling Paper, p. 73
Section 1 **First the Bad News**	90	▶ Describe the major types of pollution. ▶ Distinguish between renewable and nonrenewable resources. ▶ Explain how habitat destruction affects organisms. ▶ Explain the impact of human population growth. UCP 3, SAI 1, SPSP 1–5, HNS 3, LS 3a, 4d	**Demonstration,** Crude Oil Spills and Seabird Eggs, p. 79
Section 2 **The Good News: Solutions**	90	▶ Explain the importance of conservation. ▶ Describe the three Rs and their importance. ▶ Explain how habitats can be protected. ▶ List ways you can help protect the Earth. UCP 3, 4, SAI 1, ST 2, SPSP 1, 2, 4, 5, HNS 2, LS 5c; Labs SAI 1, 2, HNS 2, UCP 2, SPSP 4	**Demonstration,** Friendly Cleaner, p. 82 in ATE **QuickLab,** Trash Check, p. 86 **Interactive Explorations CD-ROM,** Moose Malady *A **Worksheet** is also available in the **Interactive Explorations Teacher's Edition.*** **Discovery Lab,** Biodiversity–What a Disturbing Thought! p. 134 **Datasheets for LabBook,** Biodiversity–What a Disturbing Thought! **Skill Builder,** Deciding About Environmental Issues, p. 88 **Datasheets for LabBook,** Deciding About Environmental Issues **EcoLabs & Field Activities,** A Filter with Culture **Long-Term Projects & Research Ideas,** Let's Talk Trash

*See page **T23** for a complete correlation of this book with the*

NATIONAL SCIENCE EDUCATION STANDARDS.

TECHNOLOGY RESOURCES

 Guided Reading Audio CD
English or Spanish, Chapter 4

 One-Stop Planner CD-ROM with Test Generator

 Interactive Explorations CD-ROM
CD 2, Exploration 3, Moose Malady

 Science Discovery Videodiscs
Science Sleuths: Dead Fish on Union Lake

CNN. Multicultural Connections, Thailand Tire Furniture, Segment 12

Eye on the Environment, Smog Problems in Mexico, Segment 11

Scientists in Action, Forming the Future of Energy Efficiency, Segment 7
Tracking Mercury in the Everglades, Segment 15

CLASSROOM WORKSHEETS, TRANSPARENCIES, AND RESOURCES	SCIENCE INTEGRATION AND CONNECTIONS	REVIEW AND ASSESSMENT
Directed Reading Worksheet **Science Puzzlers, Twisters & Teasers**	**Careers:** Biologist—Dagmar Werner, p. 94	
Directed Reading Worksheet, Section 1 **Transparency 165,** The Greenhouse Effect **Math Skills for Science Worksheet,** A Formula for SI Catch-up **Math Skills for Science Worksheet,** Rain-Forest Math	**Cross-Disciplinary Focus,** p. 75 in ATE **Chemistry Connection,** p. 76 **Math and More,** p. 76 in ATE **Connect to Earth Science,** p. 76 in ATE **MathBreak,** Water Depletion, p. 77 **Math and More,** p. 78 in ATE **Apply,** p. 80	**Section Review,** p. 76 **Self-Check,** p. 77 **Section Review,** p. 80 **Quiz,** p. 80 in ATE **Alternative Assessment,** p. 80 in ATE
Transparency 77, Practicing Conservation **Directed Reading Worksheet,** Section 2 **Reinforcement Worksheet,** It's "R" Planet! **Critical Thinking Worksheet,** Bud Kindfellow Has a Plan	**Math and More,** p. 82 in ATE **Connect to Chemistry,** p. 82 in ATE **Multicultural Connection,** p. 85 in ATE **Cross-Disciplinary Focus,** p. 86 in ATE **Scientific Debate:** Where Should the Wolves Roam? p. 95	**Self-Check,** p. 83 **Section Review,** p. 84 **Section Review,** p. 87 **Quiz,** p. 87 in ATE **Alternative Assessment,** p. 87 in ATE

  internet**connect**

 Holt, Rinehart and Winston On-line Resources

go.hrw.com

For worksheets and other teaching aids related to this chapter, visit the HRW Web site and type in the keyword: **HSTENV**

SC*LINKS* NSTA **National Science Teachers Association**

www.scilinks.org

Encourage students to use the *sci*LINKS numbers listed in the internet connect boxes to access information and resources on the **NSTA** Web site.

END-OF-CHAPTER REVIEW AND ASSESSMENT

Chapter Review in Study Guide
Vocabulary and Notes in Study Guide
Chapter Tests with Performance-Based Assessment, Chapter 4 Test
Chapter Tests with Performance-Based Assessment, Performance-Based Assessment 4
Concept Mapping Transparency 21

Chapter Resources & Worksheets

Visual Resources

TEACHING TRANSPARENCIES

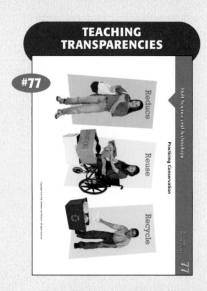

#77

TEACHING TRANSPARENCIES

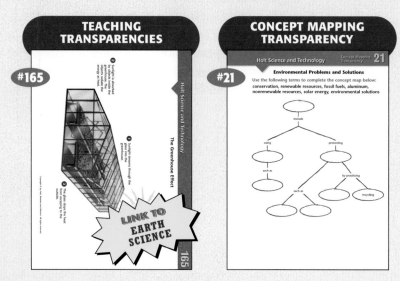

#165

CONCEPT MAPPING TRANSPARENCY

#21

Holt Science and Technology

Environmental Problems and Solutions

Use the following terms to complete the concept map below: conservation, renewable resources, fossil fuels, aluminum, nonrenewable resources, solar energy, environmental solutions

Meeting Individual Needs

DIRECTED READING

REINFORCEMENT & VOCABULARY REVIEW

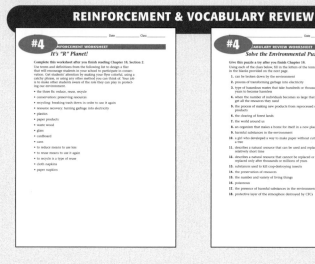

SCIENCE PUZZLERS, TWISTERS & TEASERS

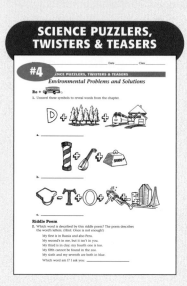

Review & Assessment

STUDY GUIDE

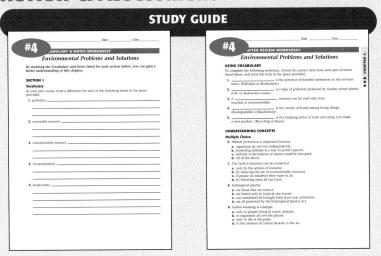

#4 VOCABULARY & NOTES WORKSHEET
Environmental Problems and Solutions

By studying the Vocabulary and Notes listed for each section below, you can gain a better understanding of this chapter.

SECTION 1
Vocabulary
In your own words, write a definition for each of the following terms in the space provided.

1. pollution

2. renewable resource

3. nonrenewable resource

4. overpopulation

5. biodiversity

#4 CHAPTER REVIEW WORKSHEET
Environmental Problems and Solutions

USING VOCABULARY
To complete the following sentences, choose the correct term from each pair of terms listed below, and write the term in the space provided.

1. _____ is the presence of harmful substances in the environment. (Pollution or Biodiversity)
2. _____ is a type of pollution produced by nuclear power plants. (CFC or Radioactive waste)
3. A _____ resource can be used only once. (nuclear or nonrenewable)
4. _____ is the variety of forms among living things. (Biodegradable or Biodiversity)
5. _____ is the breaking down of trash and using it to make a new product. (Recycling or Reuse)

UNDERSTANDING CONCEPTS
Multiple Choice
6. Habitat protection is important because
   a. organisms do not live independently.
   b. protecting habitats is a way to protect species.
   c. without it the balance of nature could be disrupted.
   d. All of the above
7. The Earth's resources can be conserved
   a. only by the actions of industry.
   b. by reducing the use of nonrenewable resources.
   c. if people do whatever they want to do.
   d. by throwing away all our trash.
8. Endangered species
   a. are those that are extinct.
   b. are found only in tropical rain forests.
   c. can sometimes be brought back from near extinction.
   d. are all protected by the Endangered Species Act.
9. Global warming is a danger
   a. only to people living in warm climates.
   b. to organisms all over the planet.
   c. only to life at the poles.
   d. to the amount of carbon dioxide in the air.

CHAPTER TESTS WITH PERFORMANCE-BASED ASSESSMENT

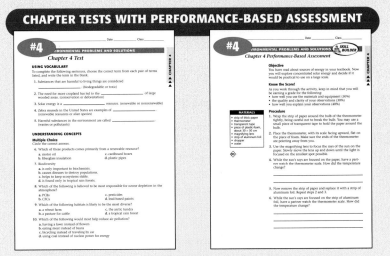

#4 ENVIRONMENTAL PROBLEMS AND SOLUTIONS
Chapter 4 Test

USING VOCABULARY
To complete the following sentences, choose the correct term from each pair of terms listed, and write the term in the blank.

1. Substances that are harmful to living things are considered _____. (biodegradable or toxic)
2. The need for more cropland has led to the _____ of large wooded areas. (conservation or deforestation)
3. Solar energy is a _____ resource. (renewable or nonrenewable)
4. Zebra mussels in the United States are examples of _____. (renewable resources or alien species)
5. Harmful substances in the environment are called _____. (wastes or pollutants)

UNDERSTANDING CONCEPTS
Multiple Choice
Circle the correct answer.
6. Which of these products comes primarily from a renewable resource?
   a. motor oil   c. cardboard boxes
   b. fiberglass insulation d. plastic pipes
7. Biodiversity
   a. is only important to biochemists.
   b. causes diseases to destroy populations.
   c. helps to keep ecosystems stable.
   d. is found only in tropical rain forests.
8. Which of the following is believed to be most responsible for ozone depletion in the atmosphere?
   a. PCBs   c. pesticides
   b. CFCs   d. lead-based paints
9. Which of the following habitats is likely to be the most diverse?
   a. a wheat farm   c. the arctic tundra
   b. a pasture for cattle d. a tropical rain forest
10. Which of the following would most help reduce air pollution?
    a. having a lawn instead of flowers
    b. eating meat instead of beans
    c. bicycling instead of traveling by car
    d. using coal instead of nuclear power for energy

#4 ENVIRONMENTAL PROBLEMS AND SOLUTIONS SKILL BUILDER
Chapter 4 Performance-Based Assessment

Objective
You have read about sources of energy in your textbook. Now you will explore concentrated solar energy and decide if it would be practical to use on a large scale.

Know the Score!
As you work through the activity, keep in mind that you will be earning a grade for the following:
- how well you use the materials and equipment (30%)
- the quality and clarity of your observations (30%)
- how well you explain your observations (40%)

MATERIALS
- strip of thick paper
- thermometer
- transparent tape
- piece of plastic-foam, about 30 × 30 cm
- magnifying lens
- strip of aluminum foil
- dropper
- water

Procedure
1. Wrap the strip of paper around the bulb of the thermometer tightly, being careful not to break the bulb. You may use a small piece of transparent tape to hold the paper around the bulb.
2. Place the thermometer, with its scale facing upward, flat on the piece of foam. Make sure the ends of the thermometer are pointing away from you.
3. Use the magnifying lens to focus the rays of the sun on the paper. Slowly move the lens up and down until the light is focused on the smallest spot possible.
4. While the sun's rays are focused on the paper, have a partner watch the thermometer scale. How did the temperature change?

5. Now remove the strip of paper and replace it with a strip of aluminum foil. Repeat steps 2 and 3.
6. While the sun's rays are focused on the strip of aluminum foil, have a partner watch the thermometer scale. How did the temperature change?

Lab Worksheets

ECOLABS & FIELD ACTIVITIES

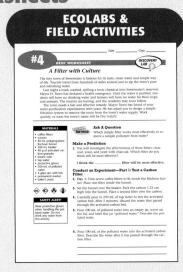

#4 STUDENT WORKSHEET DISCOVERY LAB
A Filter with Culture

The tiny town of Sweetwater is famous for its tasty, clean water and simple way of life. Tourists travel from hundreds of miles around just to sip the town's pure and refreshing water.

Last night a truck crashed, spilling a toxic chemical into Sweetwater's reservoir. Mayor H. Tuwo has declared a health emergency. Until the water is purified, residents will have no drinking water and farmers will have no water for their crops and animals. The tourists are leaving, and the residents may soon follow.

The town needs a fast and effective remedy. Mayor Tuwo has heard of your water-purification experiments with yeast. He has asked you to design a simple filtration system to remove the toxin from the town's water supply. Work quickly, or soon the town's name will be Dry Gulch!

MATERIALS
- coffee filters
- scissors
- 40 mL polypropylene Büchner funnel
- 500 mL beaker
- 40 g of activated carbon granules
- metric ruler
- tap water
- protective gloves
- 500 mL of polluted water
- 4 glass jars with lids
- permanent marker
- baker's yeast

SAFETY ALERT!
Wear protective gloves when handling the polluted water. Do not drink any water from this activity.

METHODS **Ask A Question**
Which simple filter works most effectively to remove a sample pollutant from water?

Make a Prediction
You will investigate the effectiveness of three filters: charcoal, yeast, and yeast with charcoal. Which filter do you think will be most effective?

I think the _____ filter will be most effective.

Conduct an Experiment—Part 1: Test a Carbon Filter
1. Day 1: Trim some coffee filters to fit inside the Büchner funnel. Place one filter inside the funnel.
2. Set the funnel over the beaker. Pack the carbon 1.25 cm high into the funnel. Place a second filter over the carbon.
3. Carefully pour in 250 mL of tap water to wet the activated carbon bed. After 5 minutes, discard the water that passed through the activated carbon bed.
4. Pour 100 mL of polluted water into an empty jar, screw on the lid, and label this jar "polluted water." Describe the polluted water.

6. Pour 100 mL of the polluted water into the activated carbon filter. Describe the water after it has passed through the carbon filter.

LONG-TERM PROJECTS & RESEARCH IDEAS

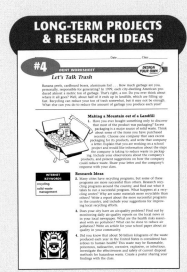

#4 STUDENT WORKSHEET DESIGN YOUR OWN
Let's Talk Trash

Banana peels, cardboard boxes, aluminum foil . . . How much garbage are you, personally, responsible for generating? In 1999, each city-dwelling American produced almost a metric ton of garbage. That's right, a ton. Do you ever think about where it all goes? Well, about half of it ends up in landfills, which are filling up fast. Recycling can reduce your ton of trash somewhat, but it may not be enough. What else can you do to reduce the amount of garbage you produce each year?

Making a Mountain out of a Landfill
1. Have you ever bought something only to discover that most of the product was packaging? Excess packaging is a major source of solid waste. Think about some of the items you have purchased recently. Choose one company that uses excess packaging for its products, and write that company a letter. Explain that you are working on a school project and would like information about the steps the company is taking to reduce waste in packaging. Include your observations about the company's products, and present suggestions on how the company could reduce waste. Share your letter and the company's response with your class.

Research Ideas
2. Many cities have recycling programs, but some of these programs are more successful than others. Research recycling programs around the country, and find out what it takes to run a successful program. What happens at a recycling center? Why are some materials more recyclable than others? Write a report about the more successful programs in the country, and include your suggestions for improving local recycling efforts.

INTERNET KEYWORDS
recycling
solid waste
management

3. Does your city have an air-quality problem? Find out by monitoring daily air-quality reports on the local news or in your local newspaper. What are the health risks associated with air pollution? What can be done to reduce air pollution? Write an article for your school paper about air quality in your community.

4. Did you know that about 50 billion kilograms of the waste produced each year in the United States is considered hazardous to human health? This waste may be flammable, poisonous, radioactive, corrosive, explosive, or infectious. Investigate the effectiveness and safety of current disposal methods for hazardous waste. Create a poster sharing your findings with the class.

DATASHEETS FOR LABBOOK

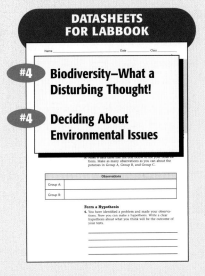

#4 **Biodiversity—What a Disturbing Thought!**

#4 **Deciding About Environmental Issues**

Make a data table like the one below to list the your country observations. Make as many observations as you can about the potatoes in Group A, Group B, and Group C.

Observations
Group A:
Group B:

Form a Hypothesis
4. You have identified a problem and made your observations. Now you can make a hypothesis. Write a clear hypothesis about what you think will be the outcome of your tests.

Applications & Extensions

CRITICAL THINKING & PROBLEM SOLVING

#4 CRITICAL THINKING WORKSHEET
Bud Kindfellow Has a Plan

Look at this article from the May 21, 2016, issue of the *Daily Post Times*.

Folk singer and billionaire Bud Kindfellow is ready to put his money where his mouth is. After 4 years of singing about the wasteful, polluting ways of modern society, Kindfellow has purchased the Pacific island of Fez. He plans to develop the island as a real-world example of "clean, sustainable living."

"The most hideous threat to our global self-being is noise pollution," says Kindfellow, "and that's the first problem we're going to solve on Fez. There will be no jackhammers, and no airplanes taking off, and all the cars will have their horns removed. The only music on the island will be my own soft, relaxing folk songs, turned off, and all the cars will be all natural."

When asked how he plans to generate electricity for the island, Kindfellow replied, "Coal burns dirty, man. Buildings on Fez will run on pure, clean nuclear

USEFUL TERMS
sustainable able to support life indefinitely with a high standard of living and health

power. In the long run, it's cheaper than going solar." Every car on Fez will be powered by 4,000 AA batteries to eliminate harmful pollutants. "That's so we don't have to deal with global warming," adds the singer.

When asked how the island's settlers will feed themselves, Kindfellow stated, "Oh man, there are tons of wild pigs there, which we can hunt for meat. And if those ever run out, we have plans to ship in some wild turkeys from South Texas. The food on Fez will be all natural."

To prevent pollution, all industrial activity will be banned from the island. Kindfellow's estimated $200 billion fortune is expected to support the island and its inhabitants for approximately 5 years.

Distinguishing Between Fact and Opinion?
1. Where in the article does Mr. Kindfellow offer an opinion?

Thinking Logically
2. Will powering cars with disposable batteries instead of fossil fuels protect the residents of Fez from global warming? Explain.

MULTICULTURAL CONNECTIONS

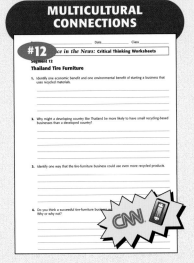

#12 Science in the News: Critical Thinking Worksheets
Segment 12
Thailand Tire Furniture

1. Identify one economic benefit and one environmental benefit of starting a business that uses recycled materials.

2. Why might a developing country like Thailand be more likely to have small recycling-based businesses than a developed country?

3. Identify one way that the tire-furniture business could use even more recycled products.

4. Do you think a successful tire-furniture business will help the island? Why or why not?

EYE ON THE ENVIRONMENT

#11 Science in the News: Critical Thinking Worksheets
Segment 11
Smog Problems in Mexico

1. Identify the natural phenomenon discussed in the video that contributes to air pollution.

2. Name the two greatest sources of pollutants in Mexico City and other industrial cities.

SCIENTISTS IN ACTION

#7 Science in the News: Critical Thinking Worksheets
Segment 7
Forming the Future of Energy Efficiency

#15 Scientists agree that efficiency is key to solving the global warming problem.

2. What incentives could home owners and businesses use to become more energy efficient?

INTERACTIVE EXPLORATIONS

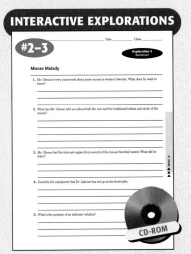

#2–3 Exploration 3 Worksheet

Moose Malady

1. Mr. Oleson is very concerned about some moose in western Sweden. What does he want to know?

2. Mr. Oleson told you about both the new and the traditional habitat and niche of the moose?

3. Mr. Oleson had the internal organs from several of the moose that died tested. What did he learn?

4. Describe the equipment that Dr. Labcoat has set up on the front table.

5. What is the purpose of an indicator solution?

CD-ROM

SECTION 1

First the Bad News

▶ Rachel Carson (1907–1964)

Rachel Carson was an American biologist and environmentalist. Her interest in nature and wildlife can be traced to her childhood in the western Pennsylvania countryside. In 1932, she received a masters degree from Johns Hopkins University and later obtained a job as an aquatic biologist with the U.S Bureau of Fisheries. She supplemented her income by writing articles about marine life. In 1941, she published her first book, *Under the Sea-Wind*—a vivid depiction of life in the oceans. Her second book, *The Sea Around Us,* was published in 1951 and was extremely well received. It became a best-seller, won the National Book Award, and established Carson as an outstanding science writer.

- Carson was apprehensive about the use of the pesticide DDT as early as 1945, but at that time few other people shared her concern. By the late 1950s, DDT use had increased enormously. It was routinely used on crops to control pests and (rather ineffectively) for mosquito control, but unfortunately it also killed nontargeted animals, such as birds, fish, crabs, grasshoppers, and bees.

- Carson began to research pesticides and turned up alarming evidence that pesticides contaminated soil, water, and air; became increasingly concentrated as they moved  up the food chain; and that such chemicals are so stable that they persist as toxins for a very long time. In the meantime, chemical companies were zealously promoting the "miracles" worked by pesticides such as DDT. Carson compiled her evidence against chemical pesticides in her book *Silent Spring* (1962). The book's release sparked an immediate controversy. Chemical companies were Carson's most vociferous opponents.

- Many scientists, however, praised Carson's book. The American public took Carson's warnings to heart, and by the end of 1962, more than 40 bills concerning pesticide regulations had been introduced in state legislatures. The formation of the Environmental Protection Agency, in 1970, was due partly to Carson's efforts. The United States began to phase out the use of DDT in 1972.

IS THAT A FACT!

- ▶ Upon the release of *Silent Spring,* representatives of chemical manufacturers did not shrink from attacking Carson personally. Carson was accused of being a "high priestess of nature," and her book was said to be part of a "communist plot" to destroy the economies of noncommunist countries.

▶ The Problems with Landfills

Landfills take up valuable land, smell bad, and attract pests. They are responsible for a host of less-obvious but equally troublesome problems as well. Decades ago, little thought was given to landfills; they were established wherever cheap or seemingly "useless" land existed. For years, household trash—some of it containing toxic chemicals, such as those found in household cleaners, paints, bug sprays, plastics, motor oil, and batteries—was plowed into landfills.

- Over time, rainwater percolated through the mountains of garbage, picking up chemical and bacterial wastes and dissolved metals. The rainwater forms a poisonous concoction known as leachate. Leachate can seep into ground water and can eventually contaminate drinking water.

IS THAT A FACT!

- ▶ In March 1987, a tugboat hauling a barge loaded with more than 3,000 tons of garbage from New York City headed to North Carolina. Officials in North Carolina turned the boat away because they feared that the trash might be hazardous. The boat traveled from port to port along the Atlantic seaboard for 2 months, and all requests to dump its load were denied. It finally returned to New York. After a 3-month delay, the refuse was incinerated in Brooklyn.

SECTION 2

The Good News: Solutions

▶ John Muir (1838-1914)

Scottish-born naturalist John Muir is often considered the father of the conservation movement in the United States. When Muir was 11, he and his family moved to the United States and settled on a farm in Wisconsin. Muir attended the University of Wisconsin for 2.5 years. In 1867, he set out to walk from Indianapolis to the Gulf of Mexico, an experience he described in his classic book, *A Thousand-Mile Walk to the Gulf* (1916). The walk was a turning point in Muir's life; he moved to Yosemite Valley, in California, and devoted himself to the study of the wilderness areas of the West.

• Muir wrote a number of natural-history articles for national magazines that opened the eyes of many to the wonders of nature. In 1876, he proposed to the federal government that it take measures to preserve forests. His efforts played a key role in the establishment of Sequoia and Yosemite National Parks in 1890.

• Two years later, Muir founded the Sierra Club, whose purpose was to "explore, enjoy, and render accessible the mountain regions of the Pacific Coast." Today the Sierra Club is a national organization with a much broader goal that includes the exploration and enjoyment of wilderness as well as the responsible use, protection, and restoration of the world's ecosystems.

▶ A Heritage of Reuse and Recycling

Before the 1950s in the United States, recycling and reuse were routine practices. Relatively few disposable items existed, and most products were not wrapped and sealed in multiple layers of packaging. A few items, such as sugar and flour, were packaged, but people either reused the large cotton sacks those items came in or made the sacks into clothing. Some of the sacks had decorative prints, encouraging their use as clothing.

IS THAT A FACT!

▪ In 1955, *Life* magazine published an article on the new "throwaway" lifestyle and how it liberated Americans from tedious cleanup chores. Among the noteworthy items the article highlighted were disposable curtains, hunting decoys, and barbecue grills.

▪ The conventional recycling of paper involves the use of harsh de-inking chemicals and requires that the paper be repulped and reprocessed into new paper. Most paper can be recycled only about three times before its cellulose fibers lose their integrity. Some paper is more suitable for recycling than others. In 1996, a mechanical engineer named Sameer Madanshetty devised a very gentle, chemical-free method that uses focused sound waves to "explode" ink off of paper. When paper is placed in water, tiny bubbles form around the inked portions of the paper. Sound waves directed at the bubbles blast the bubbles and remove the ink, which can then be filtered out of the water. The paper itself is undamaged and can be dried and reused repeatedly.

Pre-Reading Questions

Students may not know the answers to these questions before reading the chapter, so accept any reasonable response.

Suggested Answers

1. Sample answer: People damage the Earth by creating harmful pollution and hazardous waste and releasing it into the environment. People reduce biodiversity by clearing forests and filling in wetlands. People also overuse nonrenewable resources and deplete the Earth of certain minerals and water sources.

2. Sample answer: People are trying to prevent further damage to the Earth by reducing human dependence on natural resources, by reusing many of the materials we would otherwise throw away, and by recycling materials that can be reprocessed for other purposes, such as aluminum, glass, plastic, and paper.

CHAPTER
4

Sections

Environmental Problems and Solutions

Pre-Reading
Questions

1. Name three ways people damage the Earth.

2. Name three ways people are trying to prevent further damage to the Earth.

72

---

internet **connect**

**HRW
On-line
Resources**

go.hrw.com

For worksheets and other teaching aids, visit the HRW Web site and type in the keyword: **HSTENV**

SCLINKS
NSTA

www.scilinks.com

Use the *sci*LINKS numbers at the end of each chapter for additional resources on the **NSTA** Web site.

Smithsonian
Institution

www.si.edu/hrw

Visit the Smithsonian Institution Web site for related on-line resources.

CNNfyi.com

www.cnnfyi.com

Visit the CNN Web site for current events coverage and classroom resources.

Sea of Red

On the shores of a tropical island, a chemical slick coats the clear water. Even though the environment may break down the chemical over time, preventing pollution is better for the environment. In this chapter, you will learn about the natural cycles that help break down substances and move them through the environment. You will also learn about problems facing the environment today and about some solutions, including actions you can take to help.

RECYCLING PAPER

In this activity, you will be making paper without cutting down trees. Instead you will be reusing paper that has already been made.

Procedure

1. Tear up **two sheets of old newspaper** into small pieces, and put them in a **blender.** Using a **beaker,** add **1 L of water.** Cover and blend until the mixture is soupy.

2. Cover the bottom of a **square pan** with **2-3 cm of water.** Place a **wire screen** in the pan. Pour 250 mL of the paper mixture onto the screen, and spread evenly.

3. Lift the screen out of the water with the paper on it. Drain excess water into the pan. Then place the screen inside a **section of newspaper.**

4. Close the newspaper, and turn it over so that the screen is on top of the paper mixture. Cover the newspaper with a **flat board.** Press on the board to squeeze out extra water.

5. Open the newspaper, and let your paper mixture dry. Use your recycled paper to write a note to a friend!

Analysis

6. In what ways is your paper like regular paper? How is it different?

7. What could you do to improve your papermaking methods?

73

START-UP
Activity

RECYCLING PAPER

MATERIALS

FOR EACH GROUP:
- 2 sheets of old newspaper
- blender
- 1 L of water
- square pan filled with water to a level of 2–3 cm
- wire screen
- section of newspaper
- flat board

Safety Caution

Remind students to review all safety cautions and icons before beginning this lab activity. Cover all work surfaces, and have students wear aprons or smocks to protect their clothing.

The paper will have to dry at least overnight; the paper may need an additional day to dry, depending on how warm and moist the air is in your classroom. To ensure the quickest drying time, have students be sure they squeeze out as much excess water as they can.

Answers to START-UP Activity

6. Answers will vary, but students should note that their paper looks similar to regular paper, and both can be used for writing or drawing. The handmade paper is not as smooth, is easier to tear, and is thicker.

7. Answers will vary. Some students might mention using less water or squeezing more water out of the mixture.

Focus

First the Bad News

In this section, students investigate the major types of pollutants and how pollution is generated. Students also distinguish between renewable and nonrenewable resources. Finally, they explore how increases in human population and human activities strain resources and threaten the habitats of other living things.

🔔 Bellringer

Ask students:

What is the difference between a renewable resource and a nonrenewable resource? Have them write two examples of each in their ScienceLog. (Renewable resources, such as pine trees, that are used at the proper rate will last forever because they grow back or replenish themselves. Nonrenewable resources, such as oil, do not replenish themselves or grow back, or they would take thousands or millions of years to replenish.)

1 Motivate

GROUP ACTIVITY

Writing In groups of four, have students come up with a list of challenges to the environment. From that list, have each group pick the four that they feel are most important and explain why they should be priorities. Afterward, have groups share their top four challenges, and write them on the board or overhead. This list could be used to generate topics for discussion, research, or action.

Terms to Learn

pollution
renewable resource
nonrenewable resource

overpopulation
biodiversity
biodegradable

What You'll Do

- ◆ Describe the major types of pollution.
- ◆ Distinguish between renewable and nonrenewable resources.
- ◆ Explain how habitat destruction affects organisms.
- ◆ Explain the impact of human population growth.

Figure 1 *The water poured into the river by this factory is polluted with chemicals and heat. The smoke contains harmful chemicals that pollute the air.*

Figure 2 *Every year, we throw away 150 million metric tons of garbage.*

74

First the Bad News

You've probably heard it before. The air is unhealthy to breathe. The water is harmful to drink. The soil is filled with poisons. The message is that Earth is sick and in great danger.

Pollution

Pollution is the presence of harmful substances in the environment. These harmful substances, known as *pollutants,* take many forms. They may be solid materials, chemicals, noise, or even heat. Often, pollutants damage or kill the plants and animals living in the affected habitat, as shown in **Figure 1.** Pollutants may also harm humans.

Piles of Garbage Americans produce more household waste than any other nation. If stacked up, the beverage cans we use in one year could reach the moon 17 times! The average American throws away 12 kg of trash a week, which usually winds up in a landfill like the one in **Figure 2.** Businesses, mines, and industries also produce large amounts of wastes.

Billions of kilograms of this waste are classified as *hazardous waste,* which means it's harmful to humans and the environment. Many industries produce hazardous wastes, including paper mills, nuclear power plants, oil refineries, and plastic and metal processing plants. Hospitals and laboratories produce hazardous medical wastes. But industry shouldn't get all the blame. Hazardous wastes also come from homes. Old cars, paints, batteries, medical wastes, and detergents all pollute the environment.

IS THAT A FACT!

The U.S. Environmental Protection Agency defines hazardous waste as "by-products of society that can pose a substantial or potential hazard to human health or the environment when improperly managed." This includes material that is toxic, corrosive, reactive, or ignitable. Examples include many waste chemicals and radioactive waste from nuclear power plants.

Where Does It All Go? Most of our household waste goes into giant landfills. Hazardous wastes are buried in landfills specially designed to contain them. However, some companies illegally dispose of their hazardous wastes by dumping them into rivers and lakes. Some wastes are burned in incinerators designed to reduce the amount of pollutants that enter the atmosphere. But if wastes are burned improperly, they add to the pollution of the air.

Chemicals Are Everywhere Chemicals are used to treat diseases. They are also used in plastics, thermometers, paints, hair sprays, and preserved foods. In fact, chemicals are everywhere. We can't get along without chemicals. Sometimes, though, we cannot get along *with* them. Chemical pesticides used to kill crop-destroying insects also pollute the soil and water. Rachel Carson, shown in **Figure 3**, wrote about the dangers of pesticides more than three decades ago.

A class of chemicals called CFCs was once used in aerosol sprays, refrigerators, and plastics. These uses of CFCs have been banned. CFCs rise high into the atmosphere and can cause the destruction of ozone. Ozone protects the Earth from harmful ultraviolet light.

Another class of chemicals, called PCBs, was once used as insulation as well as in paints, household appliances, and other products. Then scientists learned that PCBs are *toxic,* or poisonous. PCBs are now banned, but they have not gone away. They break down very slowly in the environment, and they still pollute even the most remote areas on Earth, as shown in **Figure 4.**

High-Powered Wastes Nuclear power plants produce electricity for millions of homes and businesses. They also produce *radioactive wastes,* special kinds of hazardous wastes that take hundreds or thousands of years to become harmless. These "hot" wastes can cause cancer, leukemia, and birth defects in humans. Radioactive wastes can have harmful effects on all living things.

Figure 3 *Rachel Carson's book* Silent Spring, *published in 1962, made people aware of the environmental dangers of pesticides, especially to birds.*

Figure 4 *PCBs and other pollutants have even been found in remote parts of the Arctic.*

2 Teach

READING STRATEGY

Prediction Guide Before students begin to read the section, ask them to state whether the following statements are true or false.

- The United States produces more garbage than any other nation in the world. (true)
- Noise can be a form of pollution. (true)
- All natural resources can be reused. (false)
- The destruction of animal habitats has a negative effect on animals but does not affect humans. (false)

CROSS-DISCIPLINARY FOCUS

History Rules about garbage disposal have been on the books since at least 320 B.C., when residents of Athens, Greece, were prohibited from disposing of their garbage within city limits. During the Middle Ages, people in Europe commonly tossed their garbage into the streets, which attracted rodents and pests that often carried disease, such as bubonic plague.

Directed Reading Worksheet Section 1

internet**connect**

SCiLINKS
NSTA

TOPIC: Air Pollution
GO TO: www.scilinks.org
*sci*LINKS NUMBER: HSTL505

Science Bloopers

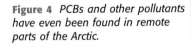

Can you imagine children going to school on top of a hazardous waste site? That's what happened in Niagara Falls, New York, in 1954 when a school and neighborhood were built on land covering an old dumping site for a chemical company. Kids played in and around pools of black muck. Parents noticed that their children were burned by the chemicals. Dogs lost their fur. Dark, smelly substances seeped into homes after rainfalls. In 1978, the state ordered the evacuation of 235 families living near the toxic site.

MATH and MORE

On average, for each mile (1.6 km) a car travels, 0.36 kg of carbon dioxide is added to the atmosphere.

For 2 weeks, have students keep a log of how many miles their family travels by car. At the end of 2 weeks, have students figure out how many total kilograms of carbon dioxide their family added to the atmosphere.

CONNECT TO
EARTH SCIENCE

Only 20 percent of the radiation that enters the Earth's atmosphere is absorbed by gases in the atmosphere and transferred in the form of heat. But these gases capture heat in other ways. When land and water absorb radiation, their molecules move faster, increasing their temperature. This energy is transferred to gas molecules in the atmosphere before it can escape into space. As a result, the atmosphere warms up. The Earth's heating process, in which the gases in the atmosphere absorb radiation and transfer the energy in the form of heat, is known as the greenhouse effect. The Earth's atmosphere works much like a greenhouse, as shown in Teaching Transparency 165, "The Greenhouse Effect."

 Teaching Transparency 165 "The Greenhouse Effect" *LINK TO EARTH SCIENCE*

 **Chemistry**
C O N N E C T I O N

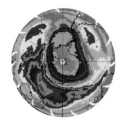

Ozone in the stratosphere absorbs most of the ultraviolet light that comes from the sun. Ozone is destroyed by CFCs. This image of the hole in the ozone layer (the gray area in the center) was taken in 1998.

Exposure to high levels of ultraviolet light can lead to blindness, rapid skin aging, skin cancer, and a weakened immune system.

internetconnect

SCI**LINKS**
NSTA

TOPIC: Air Pollution
GO TO: www.scilinks.org
*sci***LINKS NUMBER:** HSTL505

Too Much Heat The Earth is surrounded by a mixture of gases, including carbon dioxide, that make up the atmosphere. The atmosphere acts as a protective blanket, keeping the Earth warm enough for life to exist. Since the late 1800s, however, the amount of carbon dioxide in the air has increased by 25 percent. Carbon dioxide and certain pollutants in the air act like a greenhouse. Most scientists think the increase in carbon dioxide and other pollutants has caused a significant increase in global temperatures. If the temperatures continue to rise, the polar icecaps could melt, raising the level of the world's oceans. Some scientists think the sea level could rise 10 cm to 1.2 m by the year 2100. A 1 m rise would flood coastal areas, pollute underground water supplies, and cause present shorelines to disappear.

It's Way Too Noisy! Some pollutants affect the senses. These include bad odors and loud noises. Too much noise is not just annoying; it affects the ability to hear and think. If construction workers and others who work in noisy environments do not protect their ears, they can slowly lose their hearing. The students shown in **Figure 5** are listening to music at a sensible volume so that their hearing will not be damaged.

Figure 5 *Listening to music at a sensible volume will help prevent hearing loss.*

SECTION REVIEW

1. Describe two ways pollution can be harmful.

2. Explain how loud noise can be considered pollution.

3. **Applying Concepts** Explain how each of the following can help people but harm the environment: hospitals, refrigerators, and road construction.

▼ *Answers to Section Review*

1. Some pollutants are eyesores, such as open landfills. Some pollutants are poisonous to the environment and to living things.

2. Loud noise affects the ability to hear and think and can even permanently damage hearing.

3. hospitals: provide health care and generate medical waste; refrigerators: keep our food fresh and use a polluting chemical as a refrigerant; road construction: helps people travel easily from place to place and destroys habitat

Resource Depletion

Another problem for our environment is that we are using up, or depleting, natural resources. Some of the Earth's resources are renewable, but others are nonrenewable. A **renewable resource** is one that can be used again and again or has an unlimited supply. Fresh water and solar energy are renewable resources, as are some kinds of trees. A **nonrenewable resource** is one that can be used only once. Most minerals are nonrenewable. Fossil fuels, such as oil and coal, are also nonrenewable resources.

Some nonrenewable resources, such as petroleum, are probably not in danger of running out in your lifetime. But we use more and more nonrenewable resources every year, and they cannot last forever. Plus, the removal of some materials from the Earth carries a high price tag in the form of oil spills, loss of habitat, and damage from mining, as shown in **Figure 6**.

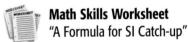

MATH BREAK

Water Depletion

An underground water supply has a depth of 200 m of water. Water seeps in at the rate of 4 cm/year. Water is pumped out at the rate of 1 m/year. How long will this water supply last?

To find the net water loss from an underground water supply, subtract the amount that seeps into the water supply from the amount removed from the water supply.

How long will the water supply last if water seeps in at the rate of 10 cm/year and is removed at the rate of 10 cm/year?

Figure 6 *This area has been mined for coal using a method called strip mining.*

Nonrenewable or Renewable? Some resources once thought to be renewable are becoming nonrenewable. Ecosystems, such as tropical rain forests, are being polluted and destroyed, resulting in huge losses of habitat. Around the world, rich soil is being eroded away and polluted. A few centimeters of soil takes thousands of years to form and can be washed away in less than a year. Underground water needed for drinking and irrigation is used faster than it is replaced. Several centimeters of water may seep into an underground source each year, but in the same amount of time, *meters* of water are being pumped out.

✓ Self-Check

1. In what ways do you use nonrenewable resources?

2. Why would it not be a good idea to use up a nonrenewable resource?

(See page 168 to check your answers.)

77

Answers to Self-Check

1. We use nonrenewable resources when we burn fossil fuels while driving or riding in a car or burning coal for heat. When we use minerals that are mined, we are using a nonrenewable resource. Pumping ground water is another use of a nonrenewable resource, if the water is used faster than it is replenished.

2. If a nonrenewable resource is used up, we can no longer rely on that resource. Certain oil and coal deposits have been building since life began on the planet. It may take hundreds of years to replace a mature forest that can be cut in a day.

While Earth's human population is expected to continue to grow rapidly, not every country in the world is expected to grow at the same rate. The growth rates of some countries, such as Japan and Sweden, are expected to shrink. In 1998, there were 5.9 billion people in the world. If current trends continue, there will be 9.4 billion in 2050, almost double. You can quickly estimate the number of years it will take for a population to double by dividing 70 by the population's current growth rate (GR). Have students find the doubling time for the following regions:

Northern Europe, GR = 0.2
North America, GR = 0.7
Africa, GR = 2.9
Asia, GR = 1.8
Latin America, GR = 2.2

Northern Europe
70 ÷ 0.2 GR = 350 years

North America
70 ÷ 0.7 GR = 100 years

Africa 70 ÷ 2.9 GR = 24 years

Asia 70 ÷ 1.8 GR = 39 years

Latin America
70 ÷ 2.2 GR = 32 years

internetconnect

SCI**LINKS**
NSTA

TOPIC: Population Growth
GO TO: www.scilinks.org
*sci*LINKS NUMBER: HSTL515

Figure 7 *The zebra mussel is an alien invader that is clogging water treatment plants in the Great Lakes region.*

Figure 8 *The purple loosestrife from Europe is choking out natural vegetation in North America.*

78

Alien Species

People are constantly on the move. Without knowing it, we take along passengers. Boats, airplanes, and cars carry plant seeds, animal eggs, and adult organisms from one part of the world to another. An organism that makes a home for itself in a new place is an *alien*. One reason alien species often thrive in foreign lands is that they are free from the predators in their native habitats.

Alien species often become pests and drive out native species. The zebra mussel, shown in **Figure 7,** hitched a ride on ships sailing from Europe to the United States in the 1980s. The purple loosestrife, shown in **Figure 8,** arrived long ago from Europe. Today it is crowding out native vegetation and threatening rare plant species in much of North America. Many organisms, such as the dandelion, are so common and have been here so long that it is easy to forget they don't belong.

Human Population Growth

In 1800, there were 1 billion people on Earth. In 1990, there were 5.2 billion. By 2100, there may be 14 billion. Today, one out of ten people goes to bed hungry every night, and millions die each year from hunger-related causes. Some people believe that the human population is already too high for the Earth to support.

More people require more resources, and the human population is growing rapidly. **Overpopulation** occurs when the number of individuals becomes so large that they can't get all the food, water, and other resources they need on an ongoing basis.

Figure 9 shows that it took most of human history for the human population to reach 1 billion. Will the planet be able to support 14 billion people?

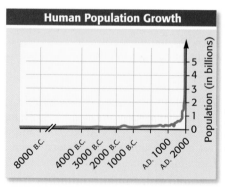

Human Population Growth

Population (in billions)
5
4
3
2
1
0

8000 B.C. 4000 B.C. 3000 B.C. 2000 B.C. 1000 B.C. A.D. 1000 A.D. 2000

Figure 9 *The Earth's human population is now doubling every few decades.*

Science Bloopers

During the 1930s, sometimes called the Dust Bowl years, soil conservation scientists in the southern United States recommended planting kudzu to hold soil in place. This fast-growing vine is native to China and Japan and can grow 18 m per season. Frosts kill kudzu in Asia, keeping its growth in check. In the South, where heavy frosts are infrequent, kudzu can grow all year. By the 1960s, kudzu had spread over trees and other plants throughout the South. Eradicating the vine is difficult because a single plant may have 50 or more vines attached to a giant root, which can weigh as much as 181 kg!

Habitat Destruction

The term **biodiversity** means "variety of life." It refers to the many different species found in a particular habitat all across the planet.

Every habitat has its own diverse combination of occupants. Every time a bulldozer digs or a chainsaw buzzes, every time hazardous wastes are dumped, a habitat is damaged, changed, or destroyed. And every time a habitat is destroyed, biodiversity is lost.

Forests Trees give us oxygen, furniture, fuel, fruits and nuts, rubber, alcohol, paper, turpentine, pencils, and telephone poles. Once trees covered twice as much land as they do today. *Deforestation,* such as that in **Figure 10,** is the clearing of forest lands. Tropical forests are cut for mines, dams, and roads. They are also cleared for paper, fuel, and building materials. But after tropical rain forests are cleared, little can grow on the land. Tropical soil doesn't have many nutrients, so it cannot be used for farming and is often abandoned.

Wetlands Wetlands were once considered unimportant. But as you know, that's not true. Wetlands help control flooding by soaking up the water from overflowing rivers. They filter pollutants from flowing water and provide breeding grounds for animals. They help prevent soil erosion and restore underground water supplies. Yet wetlands are often drained and filled to provide land for farms, homes, and shopping malls. They are dredged to keep passages open for ships and boats. Wetland habitats can also be destroyed by pollution.

Marine Habitats Oil is a major contributor to marine habitat loss. Oil from cities and industries is sometimes dumped into the ocean. Accidental spills and waste from oil tankers add more oil to the oceans. Spilled oil contaminates both open waters and coastal habitats, as shown in **Figure 11.** All the oceans are connected, so pollutants from one ocean can be carried around the world.

Figure 10 *Temperate forests are destroyed for many of the same reasons that tropical rain forests are destroyed.*

Activity

Look around the room. How many objects can you find that are made of wood? List them in your ScienceLog. Add all the products you can think of that come from trees.

TRY at HOME

Figure 11 *Oil from the* Exxon Valdez *damaged more than 2,300 km² of the Alaskan coast.*

79

3) Extend

DEMONSTRATION

Oil Spills and Seabird Eggs

(This activity can be done as a demonstration or as a group activity.) Have groups do the following activity to investigate how oil can seep through eggshells. Give each group four hardboiled eggs and a bowl filled with 250 mL of vegetable oil that has been dyed with oil-soluble red food coloring. Students will place all the eggs in the bowl, remove one egg every 5 minutes, and shell it. Have them record how much time elapses before a shelled egg shows red coloring, indicating that the oil has permeated the shell. Have students draw some conclusions about how crude oil seeping through the shells of developing seabirds might affect the unborn birds and how it might affect future seabird populations. (The oil might kill the developing birds, or it might interfere with their proper development such that if they are born, their chance of survival would be compromised. With fewer birds born or surviving to reproduce, the species' existence could be threatened.) Sheltered English

Math Skills Worksheet
"Rain-Forest Math"

IS THAT A FACT!

The destruction of coral reef habitats, which tend to be very high in biodiversity, threatens the existence of many marine organisms. Threats to coral reef habitats include the development of coastal land and damage to reefs by boat anchors.

MISCONCEPTION ALERT

Though rain forests are lush, the soil contains very few nutrients. When organisms die in a rain forest, they decompose quickly and the nutrients that are released are quickly taken up by other organisms.

4 Close

Answer to APPLY

It is not a good idea to release 1,000 balloons because the balloons will eventually burst and fall to Earth. The plastic material of the balloons will not degrade and will be a threat to wildlife.

Quiz

1. Why is it dangerous to dispose of hazardous waste by burying it? (Buried hazardous waste can exude harmful chemicals that pollute drinking water.)

2. Why is coal considered a nonrenewable resource? (Coal is nonrenewable because it can be used only once and there is a limited supply of it.)

3. What are two ways pollution can harm people? (Possible answer: Exposure to some chemicals may cause cancer and other health problems; drinking polluted water and breathing polluted air may cause lung damage.)

ALTERNATIVE ASSESSMENT

Have groups of students paint a mural of what the neighborhood near your school might look like in the year 2100 if pollution, resource depletion, and habitat destruction continue at a rapid rate. Sheltered English

Balloons Aloft

Your town is about to celebrate its 200th birthday. A giant birthday party is planned. As part of the celebration, the town plans to release 1,000 helium balloons that say "Happy Birthday to Our Town." Why is this not a good idea? What can you do to convince town officials to change their plans?

Figure 12 *This sea bird has become entangled in a plastic six-pack holder.*

Plastics are often dumped into marine habitats. They are lightweight and float on the surface. They are not **biodegradable,** so they are not broken down by the environment. Animals, such as the bird in **Figure 12,** try to eat them and often get tangled in them and die. Dumping plastics into the ocean is against the law, but it is difficult to enforce.

Effects on Humans

Trees and sea creatures are not the only organisms affected by pollution, global warming, and habitat destruction. The damage we do to the Earth affects us too. Sometimes the effect is immediate. If you drink polluted water, you may immediately get sick or even die. But sometimes the damage is not apparent right away. Some chemicals cause cancers 20 or 30 years after a person is exposed to them. Your children or grandchildren may have to deal with depleted resources.

Anything that endangers other organisms will eventually endanger us too. Taking good care of the environment requires being concerned about what is happening right now. It also requires looking ahead to the future.

BRAIN FOOD

If humans became extinct, other organisms would go on living. But if all the insects became extinct, many plants could not reproduce. Animals would lose their food supply. The organisms we depend on, and eventually all of us, would disappear from the face of the Earth.

SECTION REVIEW

1. Why do alien species often thrive?

2. Explain how human population growth is related to pollution problems.

3. **Applying Concepts** How can the destruction of wetland habitats affect humans?

80

Answers to Section Review

1. Alien species often thrive in foreign lands because they have no natural predators.

2. The larger the human population is, the more energy and materials that are needed and the more waste that is produced. Pollution problems happen when there is more waste being produced than we can dispose of without spoiling the environment.

3. Wetlands help control flooding, filter pollutants, prevent erosion, and help restore underground water supplies.

Terms to Learn

conservation
recycling
resource recovery

What You'll Do

◆ Explain the importance of conservation.
◆ Describe the three Rs and their importance.
◆ Explain how habitats can be protected.
◆ List ways you can help protect the Earth.

The Good News: Solutions

As you've seen, the news is bad. But it isn't *all* bad. In fact, there is plenty of good news. The good news is about what people can do—and are doing—to save the Earth. It is about what *you* can do to save the Earth. Just as people are responsible for damaging the Earth, people can also take responsiblity for helping to heal and preserve the Earth.

Conservation

One major way to help save the Earth is conservation. **Conservation** is the wise use of and the preservation of natural resources. If you ride your bike to your friend's house, you conserve fuel. At the same time, you prevent air pollution. If you use organic compost instead of chemical fertilizer on your garden, you conserve the resources needed to make the fertilizer. You also prevent soil and water pollution.

Practicing conservation means using fewer natural resources. It also means reducing waste. The three Rs, shown in **Figure 13**, describe three ways to conserve resources and reduce damage to the Earth: Reduce, Reuse, and Recycle.

Figure 13 *These teenagers are observing the three Rs by using a cloth shopping bag, donating outgrown clothing to be reused, and recycling plastic.*

Reduce **Reuse** **Recycle**

SCIENCE HUMOR

Dan and Ed made a great plan,
to recycle all that they can.
They reduced lots of waste
with a bit too much haste,
and recycled their mother's new van.

 Teaching Transparency 77 "Practicing Conservation"

 **Directed Reading Worksheet** Section 2

Focus

The Good News: Solutions

In this section, students discover the importance of conservation and how the strategy "reduce, reuse, and recycle" helps conserve resources. Students also learn about the importance of maintaining biodiversity and protecting habitats. Finally, they explore specific strategies they can use to help protect the environment.

Bellringer

Have students suppose they've finished reading a magazine. In their ScienceLog, have them write down at least two things they might do that would be preferable to throwing the magazine away. (Possibilities include giving it to a friend or relative, donating it to a library or homeless shelter, using it to make a collage, or recycling it.)

1) Motivate

COOPERATIVE LEARNING

Reusing Trash Organize students in groups of five or six, and give each group a plastic grocery bag to examine. Have them devise a list of at least five ways the bag can be reused. (It can be reused for groceries, used as a waterproof covering for books or papers when it rains, cut into strips and used as ribbons, used to line a wastebasket, used to hold wet clothing or shoes, used to wrap sandwiches, or used to protect surfaces from spills.)

Have groups share their ideas with the rest of the class.
Sheltered English

DEMONSTRATION

Friendly Cleaner The following recipe is for safe, all-purpose cleaner. Write it on the chalkboard so that students may copy it if they wish.

3.78 L (1 gal) hot water

59 mL ($\frac{1}{4}$ cup) borax

59 mL ($\frac{1}{4}$ cup) vinegar

30 mL (2 tbsp) phosphate-free
liquid soap

As you prepare the solution, share the following information with students. Borax (sodium tetraborate decahydrate) is a naturally occurring mineral that consists primarily of sodium and boron and works well as a cleanser, a disinfectant, and a deodorizer. Vinegar is a weak acid and is a good degreaser. The liquid soap removes dirt and grease. All the ingredients are nontoxic and are easy on the environment. When the solution is mixed, try it out in the classroom by using it on several surfaces. Sheltered English

MATH and MORE

Poster Project A faucet that drips at a rate of one drop per second wastes about 4 L of water per day. Have pairs of students incorporate this fact into a poster reminding people to turn off taps completely.

Figure 14 These scientists are studying ways to use waste products to make biodegradable plastics.

Reduce

The most obvious way to conserve the Earth's resources is to use less. This will also help reduce pollution and wastes. Some companies have started using a variety of strategies to conserve resources. They often save money in the process.

Reducing Waste and Pollution One-third of the waste from cities and towns is packaging. To conserve resources and reduce waste, products can be wrapped in less paper and plastic. Fast foods can be wrapped in thin paper instead of large plastic containers that are not biodegradable. You can choose to take your purchases without a sack if you don't need one. Scientists, such as the ones in **Figure 14,** are working to make better biodegradable plastics.

Some companies are searching for less hazardous materials to use in making products. For example, some farmers refuse to use pesticides and chemical fertilizers. They practice organic farming. They use mulch, compost, manure, and natural pesticides. Agricultural specialists are also developing new farming techniques that are better for the environment.

Reducing Use of Nonrenewable Resources Scientists are searching for alternative sources of energy. They want to avoid burning fuels and using nuclear energy. In some parts of the world solar energy heats water and powers homes, such as those shown in **Figure 15.** Engineers are working to make solar-powered cars practical. Other scientists are investigating the use of alternative power sources, such as wind, tides, and falling water.

It's Everyone's Responsibility Using fewer resources and reducing waste is not the job of industry and agriculture alone. Individuals use plenty of manufactured products and plenty of energy. They also produce large quantities of waste. Each United States citizen produces 40 times more waste than a citizen of a developing country. Why do you think this is so? What could you do to reduce the amount of trash that you produce? Everyone can take responsibility for helping to conserve the Earth's resources.

Figure 15 Rooftop solar panels provide most of the energy used in this neighborhood in Rotterdam, Holland.

82

CONNECT TO
CHEMISTRY

Biodegradable plastics contain substances that can be broken down by microbes, such as starch. Suggest that students imagine that they work for a company that uses nondegradable plastic in the packaging of their products. Have them write a persuasive letter to the company's directors advocating a changeover to biodegradable plastics.

Reuse

Do you get hand-me-down clothes from an older sibling? Do you try to fix broken sports equipment instead of thowing it away? If so, you are helping preserve the Earth by *reusing* products.

Reusing Products Every time someone reuses a plastic bag, one less bag needs to be made, and one less bag pollutes the Earth. Every time someone uses a rechargeable battery, one less battery needs to be made, and one less battery will pollute the Earth. Reusing is an important way to conserve resources and prevent pollution.

Reusing Water About 85 percent of the water used in homes goes down the drain. Communities with water shortages are experimenting with reclaiming and reusing this waste water. Some use green plants or filter-feeding animals such as clams to clean the water. The water isn't pure enough to drink, but it is fine for watering lawns and golf courses, such as the one shown in **Figure 16.**

Recycle

Recycling is a form of reuse. **Recycling** requires breaking down trash and using it again. Sometimes recycled products are used to make the same kind of products. Sometimes they are made into different products. The park bench in **Figure 17** was made from plastic foam cups, hamburger boxes, and plastic bottles that once held detergent, yogurt, and margarine. All of the containers pictured in **Figure 18** can be easily recycled.

Figure 16 *This golf course is being watered with reclaimed water.*

> ## ✔ Self-Check
>
> 1. How can you reduce the amount of electricity you use?
> 2. List five products that can be reused easily.
>
> *(See page 168 to check your answers.)*

Figure 17 *This park bench is made of melted, remolded, and reused plastic.*

Figure 18 *These containers are examples of common household trash that can be recycled.*

83

MEETING INDIVIDUAL NEEDS

Learners Having Difficulty To help students distinguish the difference between the words *reuse* and *recycle,* point out that *reuse* refers to using a particular item again and again. You might indicate this graphically by drawing a series of horizontal arrows and drawing a lunchbox (or other reusable item) between each of the arrows to indicate that it can be used many times. Then draw the triangular, three-arrowed recycling symbol on the board. Explain that the word *recycle* usually indicates that an item undergoes a process that transforms it into another item or another version of the same item. To depict recycling, you might draw three different items made from the same material (such as a newspaper, a piece of writing paper, and a cardboard box) at the three corners of the recycling symbol.
Sheltered English

RESEARCH

Writing Find out what specific materials are recycled in your town and county. For example, many areas recycle some types of plastics but not others. If you have curbside pickup, find out where you would need to take items not picked up, such as scrap metal or compost. Share your findings with the class.
PORTFOLIO

Answers to Self-Check

1. Turn off lights, CD players, radios, and computers when leaving a room. Set thermostats a little lower in the winter (wear sweaters). Don't leave the refrigerator door open while deciding what you want.

2. Sample answers: plastic bags, rechargeable batteries, water, clothing, and toys. The difference between a reused and a recycled object is that a reused article may be cleaned but is basically unchanged. A recycled article has been broken down and re-formed into another useable product.

Prediction guide Ask students:
What is "resource recovery"?
(Hint: It's not the same as
recycling.) (Resource recovery is
the process of burning garbage to
create electricity.)

BRAIN FOOD

Provide students with the following data about the time it
takes various wastes to break
down naturally.

- paper, 2–12 months
- plastic bags, 20–30 years
- aluminum cans, 200–500
  years
- plastic rings from soft-
  drink six-packs, 450 years
- plastic-foam, never

Have students apply the
reduce or reuse strategies to
the above items. Ask if it is
better to reduce the use of
some of these items rather
than reuse them. Why or
why not?

internetconnect

SC/LINKS
NSTA

TOPIC: Recycling
GO TO: www.scilinks.org
*sci*LINKS NUMBER: HSTL520

Figure 19 *Each kind of recycled
material is sorted into its own bin
and then delivered to a recycling
plant for processing.*

Recycling Trash Plastics, paper, aluminum cans, waste wood,
glass, and cardboard are some examples of materials that can
be recycled. Every week, half a million trees are needed to
make Sunday newspapers. Recycling newspapers could save
many trees. Recycling aluminum foil and cans saves 95 percent of the energy needed to change raw ore into aluminum.
Glass makes up 8 percent of all our waste. It can be remelted
to make new bottles and jars. Lead batteries can be recycled
into new batteries.

Some cities, such as Austin, Texas, make recycling easy.
Special containers for glass, plastic, aluminum, and paper are
provided to each city customer. Each week trash to be recycled is collected in special trucks, such as the one shown in
Figure 19, at the same time other waste is collected.

Recycling Resources Waste that can be burned can also be
used to generate electricity in factories like the one shown in
Figure 20. The process of transforming garbage to electricity
is called **resource recovery.** The waste collected by all the cities
and towns in the United States could produce about the same
amount of electricity as 15 large nuclear power plants. Some
companies are beginning to do this with their own
waste. It saves them money, and it is responsible management.

Recycling is not difficult. Yet in the
United States, only about 11 percent of
the garbage is recycled. This compares
with about 30 percent in Europe and
50 percent in Japan.

Figure 20 *A waste-to-energy
plant can provide electricity to
many homes and businesses.*

SECTION REVIEW

1. Define and explain *conservation.*

2. Describe the three main ways to conserve natural
   resources.

3. **Analyzing Relationships** How does conservation of
   resources also reduce pollution and other damage to
   the Earth?

84

▼ *Answers to Section Review*

1. Conservation is the wise use and preservation of natural resources. Consuming as few
   manufactured products as possible and
   then reusing them in some way is one
   way to conserve natural resources.

2. The three main ways to conserve natural
   resources are to reduce, reuse, and recycle.

3. Conservation of resources reduces damage
   to the Earth by reducing the amount of
   resources extracted and by reducing the
   waste and pollution created to alter and
   package those resources into products
   people buy.

Maintaining Biodiversity

Imagine a forest with just one kind of tree. If a disease hits that species, the entire forest might be wiped out. Now imagine a forest with 10 different kinds of trees. If a disease hits one kind of tree, nine different species will remain. Look at **Figure 21.** This field is growing a very important crop—cotton. But it is not very diverse. For the crop to thrive, the farmer must carefully manage the crop with weedkillers, pesticides, and fertilizers. Biodiversity helps to keep communities naturally stable.

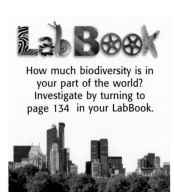

How much biodiversity is in your part of the world? Investigate by turning to page 134 in your LabBook.

Figure 21 *What could happen if a cotton disease hits this cotton field? Biodiversity is low in fields of crops like this one.*

Species variety is also important because each species makes a unique contribution to an ecosystem. In addition, many species are important to humans. They provide many things, such as foods, medicines, natural pest control, beauty, and companionship, to name just a few.

Species Protection One way to maintain biodiversity is through the protection of individual species. In the United States, the Endangered Species Act is designed to do just that. Endangered organisms are included in a special list. The law forbids activities that would damage a plant or animal on the endangered species list. It also requires the development of programs to help endangered populations recover. Some endangered species are now increasing in number, such as the California condor in **Figure 22.**

Unfortunately, the process of getting a species on the endangered list takes a long time. Many new species need to be added to the list. Many species become extinct even before they are listed!

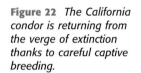

Figure 22 *The California condor is returning from the verge of extinction thanks to careful captive breeding.*

LabBook PG 134

Biodiversity—What a Disturbing Thought!

Multicultural CONNECTION

Although Costa Rica is a small country (51,000 km²), an estimated 505,660 species live there, which is about 4 percent of all living species. In 1989, the Costa Rican government set up the National Institute of Biodiversity (Instituto Nacional de Biodiversidad, or INBio) to catalog native species and to educate residents about the importance of preserving biodiversity. Some of the collecting and cataloging is done by local people who have been trained by scientists. Invite interested students to use the Internet to find out how INBio shares information about beneficial and sustainable uses for some of the unique species found in Costa Rica.

MISCONCEPTION ///ALERT\\\

In the recent past, humans caused extinction primarily by overhunting and overharvesting. Today extinction of whole groups of species is more likely to result from habitat loss and the introduction of foreign species than from hunting. Saving species today requires protecting not just the animals but the ecosystems that support them.

IS THAT A FACT!

Just outside Detroit, Michigan, looms a 50 m high former landfill once known as Mount Trashmore. From the 1970s to the mid-1990s, Mount Trashmore was used as a ski slope; it even had ski lifts! In addition, methane gas tapped from the decomposing wastes within the landfill was converted to electricity, which fulfilled the energy needs of some 2,000 households. Today the former landfill, renamed Riverview Highlands, is a recreational area offering ice skating, tubing (using recycled tractor tire tubes), and other sporting activities.

internet**connect**

TOPIC: Maintaining Biodiversity
GO TO: www.scilinks.org
*sci*LINKS **NUMBER:** HSTL525

MEETING INDIVIDUAL NEEDS

Advanced Learners There are numerous volunteer activities that seek to protect and maintain animal and plant habitats. A few activities are as follows: an organized cleanup of a beach or a riverfront; tree planting; or removing alien species from a forest, prairie, or wetland preserve. Have interested students contact local environmental organizations to find out how they can help organize and lead a group of classmates in volunteer field work.
Sheltered English

CROSS-DISCIPLINARY FOCUS

Art and Design As a habitat disappears, many birds have difficulty obtaining food and shelter. Have students or groups of students research, design, and build a simple birdhouse or bird feeder. Birdhouses and bird feeders are also a good way to make use of scrap lumber.
Sheltered English

INDEPENDENT PRACTICE

Concept Mapping
Have students make a concept map with the terms *reduce, reuse,* and *recycle* arranged around the phrase *Things I Can Do to Conserve Natural Resources.* Have them include as many ideas as they can on the map.

Figure 23 *Setting aside public lands for wildlife is one way to protect habitats.*

Habitat Protection Waiting until a species is almost extinct to begin protecting it is like waiting until your teeth are rotting to begin brushing them. Scientists want to prevent species from becoming endangered as well as from becoming extinct.

Plants, animals, and microorganisms do not live independently. Each is part of a huge interconnected web of organisms. To protect the entire web and to avoid disrupting the worldwide balance of nature, complete habitats, not just individual species, must be preserved. *All* species, not just those that are endangered, must be protected. All of the species that live in the nature preserve pictured in **Figure 23** are protected because the entire habitat is protected.

Strategies

Laws have been enacted to help conserve the Earth's environment. The purposes of such laws are listed below along with some of the ways citizens can help achieve these goals.

- **Reduce pesticide use.**
  Spray only pesticides that are targeted specifically for harmful insects. Use natural pesticides that interfere with the ways certain insects grow, develop, and live. Develop more biodegradable pesticides that will not injure birds, animals, or plants.

- **Reduce pollution.**
  Regulations prohibit the dumping of toxic substances and solid wastes into rivers, streams, lakes, and oceans and onto farmland and forests.

- **Protect habitats.**
  Conserve wetlands. Reduce deforestation. Practice logging techniques that consider the environment. Use resources at a rate that allows them to be replenished. Protect entire habitats.

- **Enforce the Endangered Species Act.**
  Speed up the process of getting endangered organisms listed.

- **Develop alternative energy sources.**
  Increase the use of solar power, wind power, and other renewable energy sources.

QuickLab

Trash Check
Keep track of all the trash you produce in one day. Classify it into groups. How much is food scraps? What might be considered a hazardous waste? What can be recycled? What can be reused? How can you reduce the amount of trash you produce?

TRY at HOME

SCIENCE HUMOR

Q: What are people who damage the habitat of endangered birds engaging in?

A: fowl play

What *You* Can Do

Reduce, reuse, recycle. Protect the Earth. These are jobs for everyone. Children as well as adults can help to save the Earth. The following list offers some suggestions for how *you* can help. How many of these things do you already do? What can you add to the list?

1. Buy things in packages that can be recycled.
2. Give away your old toys.
3. Use recycled paper.
4. Fill up both sides of a sheet of paper.
5. If you can't use permanent dishes, use paper instead of plastic-foam cups and plates.
6. Recycle glass, plastics, paper, aluminum, and batteries.
7. Don't buy anything made from an endangered animal.
8. Use rechargeable batteries.
9. Turn off lights, CD players, and computers when not in use.
10. Wear hand-me-downs.
11. Share books with friends, or use the library.
12. Walk, ride a bicycle, or use public transportation.
13. Carry a reusable cloth shopping bag to the store.
14. Use a lunch box or reuse your paper lunch bags.
15. Turn off the water while you brush your teeth.
16. Make a compost heap.
17. Buy products made from biodegradable plastic.
18. Use cloth napkins and kitchen towels.
19. Buy products with little or no packaging.
20. Repair leaking faucets.

THIS BAG IS BIODEGRADABLE

SECTION REVIEW

1. Describe why biodiversity is important.
2. Why is it important to protect entire habitats?
3. **Applying Concepts** In the list above, identify which suggestions involve reducing, reusing, or recycling. Some suggestions will involve more than one of the three Rs.

internetconnect

sci**LINKS**
NSTA

TOPIC: Recycling, Maintaining Biodiversity
GO TO: www.scilinks.org
*sci***LINKS NUMBER:** HSTL520,HSTL525

87

▼ Answers to Section Review

1. Biodiversity helps to keep biological communities naturally stable.
2. Entire habitats must be preserved and protected in order to preserve biodiversity because each organism is connected to all others in a huge web. The entire web must be protected to avoid disrupting the worldwide balance of nature.
3. 1—reduce, recycle; 2—reduce, reuse; 3—reduce, recycle; 4—reduce; 5—reduce; 6—reduce; 7—reduce; 8—reduce, reuse; 9—reduce; 10—reduce, reuse; 11—reduce, reuse; 12—reduce, reuse; 13—reduce, reuse; 14—reduce, reuse; 15—recycle; 16—recycle; 17—reduce; 18—reduce; 19—reduce; 20—reduce

Deciding About Environmental Issues
Teacher's Notes

Time Required
One 45-minute class period

Lab Ratings

EASY ————————————→ HARD

TEACHER PREP   🧪
STUDENT SET-UP   🧪🧪
CONCEPT LEVEL   🧪🧪🧪
CLEAN UP   🧪

MATERIALS

You may want to have students bring in articles for several days prior to doing this activity. You might also want to limit students' choices to those that are age-appropriate.

Lab Notes

This is a good lab to repeat as environmental issues appear in the news.

You may wish to combine this activity with a video that portrays an international environmental issue. Students can also be encouraged to use the Internet as a source of information.

Debra Sampson
Booker T. Washington
Middle School
Elgin, Texas

Skill Builder Lab

Deciding About Environmental Issues

You make hundreds of decisions every day. Some of them are complicated, but many of them are very simple, such as what to wear or what to eat for lunch. Deciding what to do about an environmental issue can be more difficult. Many different factors must be considered. How will a certain solution affect people's lives? How much will it cost? Is it ethically right?

In this activity, you will analyze an issue in four steps to help you make a decision about it. Find out about environmental issues that are being discussed in your area. Examine newspapers, magazines, and other publications to find out what the issues are. Choose one local issue to evaluate. For example, you could evaluate whether the city should spend the money to provide recycling bins and special trucks for picking up recyclable trash.

MATERIALS

- newspapers, magazines, and other publications containing information about environmental issues

Procedure

1. In your ScienceLog, write a statement about an environmental issue.

2. *Gather Information* Read about your issue in several publications. Summarize important facts in your ScienceLog.

3. *Consider Values* Values are the things that you consider important. Examine the diagram below. Several values are given. Which values do you think apply most to the environmental issue you are considering? Are there other values that you believe will help you make a decision about the issue? Consider at least four values in making your decision.

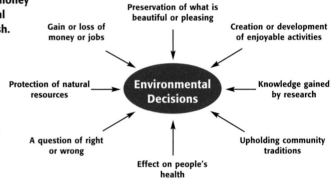

Preservation of what is beautiful or pleasing

Gain or loss of money or jobs

Creation or development of enjoyable activities

Protection of natural resources

Environmental Decisions

Knowledge gained by research

A question of right or wrong

Effect on people's health

Upholding community traditions

88

Datasheets for LabBook

Consequences				
Types of consequences	**Value 1:**	**Value 2:**	**Value 3:**	**Value 4:**
Positive short-term consequences				
Negative short-term consequences				
Positive long-term consequences				
Negative long-term consequences				

DO NOT WRITE IN BOOK

4 *Explore Consequences* Consequences are the things that result from a certain course of action. In your ScienceLog, create a table similar to the one above. Use your table to organize your thoughts about consequences related to your environmental issue. List your values at the top. Fill in each space with the consequences of an action related to each of your values.

5 *Make a Decision* Thoroughly consider all of the consequences you have recorded in your table. Evaluate how important each consequence is. Make a decision about what course of action you would choose on the issue.

Analysis

6 In your evaluation, did you consider short-term consequences or long-term consequences to be more important? Why?

7 Which value or values had the greatest influence on your final decision? Explain your reasoning.

Going Further

Compare your table with your classmates' tables. Did you all make the same decision about a similar issue? If not, form teams and organize a formal classroom debate on a specific environmental issue.

89

Answers

6. Either short- or long-term consequences can be more relevant, depending on the issue.

7. Answers will vary according to students' perspectives.

Going Further

Encourage students to narrow their topic to a single aspect of an issue, such as the importance of aesthetic value in the preservation of natural areas.

 Science Skills Worksheet "Thinking Objectively"

 Science Skills Worksheet "Working with Hypotheses"

Chapter Highlights

Chapter Highlights

VOCABULARY DEFINITIONS

SECTION 1

pollution the presence of harmful substances in the environment

renewable resource a natural resource that can be used and replaced over a relatively short time

nonrenewable resource a natural resource that cannot be replaced or that can be replaced only over thousands or millions of years

overpopulation a condition that occurs when the number of individuals becomes so large that there are not enough resources for all individuals

biodiversity the number and variety of living things

biodegradable capable of being broken down by the environment

---

SECTION 1

Vocabulary

pollution (p. 74)
renewable resource (p. 77)
nonrenewable resource (p. 77)
overpopulation (p. 78)
biodiversity (p. 79)
biodegradable (p. 80)

Section Notes

• The Earth is being polluted by solid wastes, hazardous chemicals, radioactive materials, noise, and heat.

• Some of the Earth's resources renew themselves, and others do not. Some of the non-renewable resources are being used up.

• Alien species often invade foreign lands, where they may thrive, become pests, and threaten native species.

• The human population is in danger of reaching numbers that the Earth cannot support.

• The Earth's habitats are being destroyed in a variety of ways, including deforestation, the filling of wetlands, and pollution.

• Deforestation may cause the extinction of species and often leaves the soil infertile.

• Air, water, and soil pollution can damage or kill animals, plants, and microorganisms.

• Humans depend on many different kinds of organisms. Pollution, global warming, habitat destruction—anything that affects other organisms will eventually affect humans too.

☑ Skills Check

Math Concepts

NET WATER LOSS Suppose that water seeps into an underground water supply at the rate of 10 cm/year. The underground water supply is 100 m deep, but it is being pumped out at about 2 m/year. How long will the water last?

First convert all measurements to centimeters.

(100 m = 10,000 cm; 2 m = 200 cm)

Then find the net loss of water per year.

200 cm – 10 cm = 190 cm (net loss per year)

Now divide the depth of the underground water supply by the net loss per year to find out how many years this water supply will last.

10,000 cm ÷ 190 cm = 52.6 years

Visual Understanding

THINGS YOU CAN DO Obviously, the strategies listed on page 86 to help preserve the Earth's habitats are strategies that scientists and other professionals are developing. To help you understand some of the things that you can do now, review the list on page 87.

90

Lab and Activity Highlights

Deciding About Environmental Issues `PG 88`

Biodiversity—What a Disturbing Thought! `PG 134`

 Datasheets for LabBook (blackline masters for these labs)

SECTION 2

Vocabulary

conservation *(p. 81)*

recycling *(p. 83)*

resource recovery *(p. 84)*

Section Notes

- Conservation is the wise use of and preservation of the Earth's natural resources. By practicing conservation, people can reduce pollution and ensure that resources will be available to people in the future.

- Conservation involves the three Rs: Reduce, Reuse, and Recycle. Reducing means using fewer resources to begin with. Reusing means using materials and products over and over. Recycling involves breaking down used products and making them into new ones.

- Biodiversity is the variety of life on Earth. It is vital for maintaining stable, healthy, and functioning ecosystems.

- Habitats can be protected by using fewer pesticides, reducing pollution, avoiding habitat destruction, protecting species, and using alternative renewable sources of energy.

- Everyone can help to save the Earth by practicing the three Rs in their daily life.

Labs

Biodiversity—What a Disturbing Thought! *(p. 134)*

conservation the wise use of and preservation of natural resources

recycling the process of making new products from reprocessed used products

resource recovery the process of transforming into useable products things normally thrown away

Vocabulary Review Worksheet

Blackline masters of these Chapter Highlights can be found in the **Study Guide.**

internet **connect**

GO TO: go.hrw.com

Visit the **HRW** Web site for a variety of learning tools related to this chapter. Just type in the keyword:

KEYWORD: HSTENV

SCi**LINKS**™

N S T A

GO TO: www.scilinks.org

Visit the **National Science Teachers Association** on-line Web site for Internet resources related to this chapter. Just type in the *sci*LINKS number for more information about the topic:

TOPIC: Air Pollution	***sci*LINKS NUMBER:** HSTL505
TOPIC: Resource Depletion	***sci*LINKS NUMBER:** HSTL510
TOPIC: Population Growth	***sci*LINKS NUMBER:** HSTL515
TOPIC: Recycling	***sci*LINKS NUMBER:** HSTL520
TOPIC: Maintaining Biodiversity	***sci*LINKS NUMBER:** HSTL525

91

Lab and Activity Highlights

LabBank

 EcoLabs & Field Activities, A Filter with Culture

Long-Term Projects & Research Ideas, Let's Talk Trash

Interactive Explorations CD-ROM

 CD 2, Exploration 3, "Moose Malady"

Chapter Review
Answers

USING VOCABULARY

1. Pollution
2. Radioactive waste
3. nonrenewable
4. Biodiversity
5. Recycling

UNDERSTANDING CONCEPTS

Multiple Choice

6. d
7. b
8. c
9. b
10. c
11. b

Short Answer

12. ride a bicycle, recycle cans and bottles, wear secondhand clothing, use rechargeable batteries, and others

13. Alien species thrive without their native predators and often crowd out native organisms. Sometimes the habitat of a rare species is taken over by an alien species, causing the native species to become endangered.

Concept Mapping Transparency 21

Blackline masters of this Chapter Review can be found in the **Study Guide**.

Chapter Review

USING VOCABULARY

To complete the following sentences, choose the correct term from each pair of terms listed below:

1. __?__ is the presence of harmful substances in the environment. *(Pollution or Biodiversity)*

2. __?__ is a type of pollution produced by nuclear power plants. *(CFC or Radioactive waste)*

3. A __?__ resource can be used only once. *(nuclear or nonrenewable)*

4. __?__ is the variety of forms among living things. *(Biodegradable or Biodiversity)*

5. __?__ is the breaking down of trash and using it to make a new product. *(Recycling or Reuse)*

UNDERSTANDING CONCEPTS

Multiple Choice

6. Habitat protection is important because
   a. organisms do not live independently.
   b. protecting habitats is a way to protect species.
   c. without it the balance of nature could be disrupted.
   d. All of the above

7. The Earth's resources can be conserved
   a. only by the actions of industry.
   b. by reducing the use of nonrenewable resources.
   c. if people do whatever they want to do.
   d. by throwing away all our trash.

8. Endangered species
   a. are those that are extinct.
   b. are found only in tropical rain forests.
   c. can sometimes be brought back from near extinction.
   d. are all protected by the Endangered Species Act.

9. Global warming is a danger
   a. only to people living in warm climates.
   b. to organisms all over the planet.
   c. only to life at the poles.
   d. to the amount of carbon dioxide in the air.

10. Overpopulation
    a. does not occur among human beings.
    b. helps keep pollution levels down.
    c. occurs when a species cannot get all the food, water, and other resources it needs.
    d. occurs only in large cities.

11. Biodiversity
    a. is of no concern to scientists.
    b. helps to keep ecosystems stable.
    c. causes diseases to destroy populations.
    d. is found only in temperate forests.

Short Answer

12. Describe how you can help to conserve resources. Include strategies from all of the three Rs.

13. Describe the connection between alien species and endangered species.

92

Concept Mapping

14. Use the following terms to create a concept map: pollution, pollutants, CFCs, cancer, PCBs, toxic, radioactive wastes, global warming.

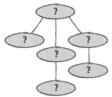

CRITICAL THINKING AND PROBLEM SOLVING

Write one or two sentences to answer the following question:

15. Suppose that the supply of fossil fuels were going to run out in 10 years. What would happen if we ran out without being prepared? What could be done to prepare for such an event?

MATH IN SCIENCE

16. If each person in a city of 150,000 throws away 12 kg of trash every week, how many metric tons of trash does the city produce per year? (There are 52 weeks in a year and 1,000 kg in a metric ton.)

INTERPRETING GRAPHICS

The illustration above shows how people in one home use natural resources.

17. Identify ways in which the people in this picture are wasting natural resources. Describe at least three examples, and tell what could be done to conserve resources.

18. Identify which resources in this picture are renewable.

19. Identify any sources of hazardous waste in this picture.

20. Explain how the girl wearing headphones is reducing pollution in the air. How could such a choice cause her harm?

Reading Check-up

Take a minute to review your answers to the Pre-Reading Questions found at the bottom of page 72. Have your answers changed? If necessary, revise your answers based on what you have learned since you began this chapter.

93

Concept Mapping

14. An answer to this exercise can be found at the front of this book.

CRITICAL THINKING AND PROBLEM SOLVING

15. We would have severe transportation problems, and most people would not be able to get to work or school. Some places would not have electricity or gas for heating, cooling, or refrigeration. Our survival might even be threatened if needed supplies, such as food and medicine, could not be transported and delivered. We could prepare for such an event by developing alternative sources of fuel, such as solar energy.

MATH IN SCIENCE

16. 93,600 metric tons per year

INTERPRETING GRAPHICS

17. The water is running while the boy is brushing his teeth. He should turn off the water while he brushes. The water sprinklers are running while it is raining. The sprinklers should be turned off; the rain will water the lawn. A boy is throwing away a plastic container that can be recycled.

18. The water is a renewable resource as long as it is replenished faster than it is used.

19. If the girl is not using rechargeable batteries in the radio, the batteries are a source of hazardous waste. The father is pouring oil from the car into the gutter, where it will be a hazardous pollutant.

20. Headphones help reduce noise pollution. If she has the volume too high, it could damage her hearing.

Background

The green iguana (*Iguana iguana*) is native to Central America, where it was used by native people as a food source. Iguana meat is tasty, is high in protein, and has a relatively low fat content.

The green iguana population in Costa Rica has declined because people have hunted iguana eggs and adults and because so much of the iguana's habitat has been destroyed. Iguanas are tree-dwelling lizards, so they require a forest habitat to survive.

Teaching Strategy

Ask students why iguana farming might provide Central Americans with environmental benefits. Lead them to conclude that since iguanas are adapted to the forest that is already there, farming them requires preserving areas of forest. Preserving forest and reforesting cleared areas in turn helps preserve fragile tropical forest soils and helps preserve the variety of other creatures that make their home in the forest.

CAREERS

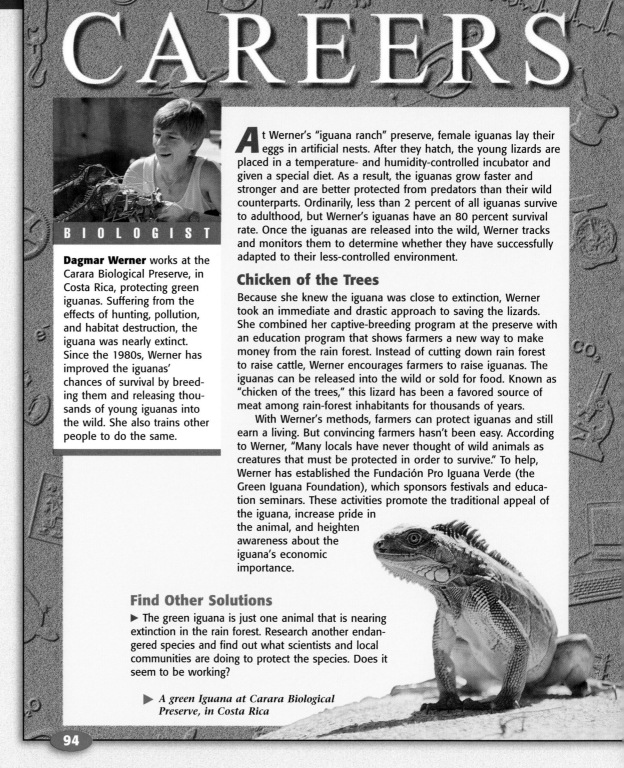

BIOLOGIST

Dagmar Werner works at the Carara Biological Preserve, in Costa Rica, protecting green iguanas. Suffering from the effects of hunting, pollution, and habitat destruction, the iguana was nearly extinct. Since the 1980s, Werner has improved the iguanas' chances of survival by breeding them and releasing thousands of young iguanas into the wild. She also trains other people to do the same.

At Werner's "iguana ranch" preserve, female iguanas lay their eggs in artificial nests. After they hatch, the young lizards are placed in a temperature- and humidity-controlled incubator and given a special diet. As a result, the iguanas grow faster and stronger and are better protected from predators than their wild counterparts. Ordinarily, less than 2 percent of all iguanas survive to adulthood, but Werner's iguanas have an 80 percent survival rate. Once the iguanas are released into the wild, Werner tracks and monitors them to determine whether they have successfully adapted to their less-controlled environment.

Chicken of the Trees

Because she knew the iguana was close to extinction, Werner took an immediate and drastic approach to saving the lizards. She combined her captive-breeding program at the preserve with an education program that shows farmers a new way to make money from the rain forest. Instead of cutting down rain forest to raise cattle, Werner encourages farmers to raise iguanas. The iguanas can be released into the wild or sold for food. Known as "chicken of the trees," this lizard has been a favored source of meat among rain-forest inhabitants for thousands of years.

With Werner's methods, farmers can protect iguanas and still earn a living. But convincing farmers hasn't been easy. According to Werner, "Many locals have never thought of wild animals as creatures that must be protected in order to survive." To help, Werner has established the Fundación Pro Iguana Verde (the Green Iguana Foundation), which sponsors festivals and education seminars. These activities promote the traditional appeal of the iguana, increase pride in the animal, and heighten awareness about the iguana's economic importance.

Find Other Solutions

▶ The green iguana is just one animal that is nearing extinction in the rain forest. Research another endangered species and find out what scientists and local communities are doing to protect the species. Does it seem to be working?

▶ *A green Iguana at Carara Biological Preserve, in Costa Rica*

94

Answer to Find Other Solutions

Students' answers will vary.

Where Should the Wolves Roam?

The U.S. Fish and Wildlife Service has listed the gray wolf as an endangered species throughout most of the United States and has devised a plan to reintroduce the wolf to Yellowstone National Park, central Idaho, and northwestern Montana. The goal is to establish a population of at least 100 wolves at each location. If the project continues as planned, wolves may be removed from the endangered species list by 2002. But some ranchers and hunters are uneasy about the plan, and some environmentalists and wolf enthusiasts think that the plan doesn't go far enough to protect wolves.

Does the Plan Risk Livestock?

Ranchers are concerned that the wolves will kill livestock. These losses could result in a tremendous financial burden to ranchers. There is a compensation program currently established that will pay ranchers if wolves kill their livestock. But this program will end if the wolf is removed from the endangered species list. Ranchers point out that the threat to their livestock will not end when the wolf is removed from the list. In fact, the threat will increase, but ranchers will no longer receive any compensation.

On the other hand, some biologists offer evidence that wolves living near areas with adequate populations of deer, elk, moose, and other prey do not attack livestock. In fact, fewer than five wolf attacks on livestock were reported between 1995 and 1997.

Are Wolves a Threat to Wildlife?

Many scientists believe that the reintroduction plan would bring these regions into ecological balance for the first time in 60 years. They believe that the wolves will eliminate old and weak elk, moose, and deer and help keep these populations from growing too large.

Hunters fear that the wolves will kill many of the game animals in these areas. They cite studies that say large game animal populations can-

◄ *A Gray Wolf in Montana*

not survive hunting by both humans and wolves. Hunting plays a significant role in the economy of the western states.

Are the People Safe?

Some people fear that wolves will attack people. However, there has never been a documented attack on humans by healthy wolves in North America. Supporters say that wolves are shy animals that prefer to keep their distance from people.

Most wolf enthusiasts admit that there are places where wolves belong and places where wolves do not belong. They believe that these reintroduction zones offer places for wolves to thrive without creating problems.

What Do You Think?

▶ Some people argue that stories about "the big, bad wolf" give the wolf its ferocious reputation. Do you think people's fears are based on myth, or do you think that the wolf is a danger to people and livestock living in the reintroduction zones? Do some research and provide examples to support your opinion.

Background

In the 1920s, the gray wolf was exterminated from much of the northwestern United States. Ranchers and federal agents killed wolves to protect livestock. Those in favor of wolf reintroduction cite biologists' claims that wolf attacks on livestock are neither as widespread nor as serious as is generally believed. Some opponents of the reintroduction plan argue that wolves should not be classified as endangered at all. According to data from biologists, there are 1,500 to 2,000 wolves in Minnesota, 6,000 to 10,000 in Alaska, and 40,000 to 50,000 in Canada. With such numbers, many people feel that the animal should not receive the special treatment given to endangered species.

Activity

Research Have interested students research the latest news on the wolf reintroduction issue and present a report to the class.

95

Answer to What Do You Think?
Students' answers might vary, but most of the fears associated with wolves are based on myths.

Chapter Organizer

CHAPTER ORGANIZATION	TIME MINUTES	OBJECTIVES	LABS, INVESTIGATIONS, AND DEMONSTRATIONS
Chapter Opener pp. 96–97	45	National Standards: SAI 1, ST 2, SPSP 5	**Start-Up Activity,** What Is the Sun's Favorite Color? p. 97
Section 1 Natural Resources	90	▶ Determine how humans use natural resources. ▶ Contrast renewable resources with nonrenewable resources. ▶ Explain how humans can conserve natural resources. SAI 1, SPSP 2, 4	
Section 2 Fossil Fuels	135	▶ Classify the different forms of fossil fuels. ▶ Explain how fossil fuels are obtained. ▶ Identify problems with fossil fuels. ▶ List ways to deal with fossil-fuel problems. SAI 1, SPSP 1, 2, 4, 5, ES 1k, 3d	**QuickLab,** Rock Sponge, p. 103 **Demonstration,** Simulating Reservoirs, p. 103 in ATE
Section 3 Alternative Resources	90	▶ Describe alternatives to the use of fossil fuels. ▶ List advantages and disadvantages of using alternative energy resources. UCP 5, ST 2, SPSP 2, 4, 5, ES 3d; Labs UCP 2, 3, SAI 1, ST 1	**Interactive Explorations CD-ROM,** The Generation Gap *A **Worksheet** is also available in the **Interactive Explorations Teacher's Edition.*** **Skill Builder,** Make a Water Wheel, p. 116 **Discovery Lab,** Power of the Sun, p. 136 **Calculator-Based Labs,** Solar Homes **Long-Term Projects & Research Ideas,** Build a City–Save a World!

*See page **T23** for a complete correlation of this book with the*

NATIONAL SCIENCE EDUCATION STANDARDS.

TECHNOLOGY RESOURCES

 Guided Reading Audio CD
English or Spanish, Chapter 5

 One-Stop Planner CD-ROM with Test Generator

 Science Discovery Videodiscs
Image and Activity Bank with Lesson Plans: Energy Resources Tour
Science Sleuths: The Energy Mystery House

 CNN. Scientists in Action, Forming the Future of Energy Efficiency, Segment 7

Science, Technology & Society, BioDiesel, Segment 6

Wind Power, Segment 14

Multicultural Connections, China's Solar Nomads, Segment 10

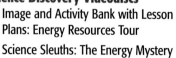 **Interactive Explorations CD-ROM**
CD 1, Exploration 6, The Generation Gap

Chapter 5 • Energy Resources

CLASSROOM WORKSHEETS, TRANSPARENCIES, AND RESOURCES	SCIENCE INTEGRATION AND CONNECTIONS	REVIEW AND ASSESSMENT
Science Puzzlers, Twisters & Teasers **Directed Reading Worksheet**		
Directed Reading Worksheet, Section 1 **Transparency 77,** Practicing Conservation **Reinforcement Worksheet,** What Are My Resources?	**Connect to Environmental Science,** p. 98 in ATE **Real-World Connection,** p. 99 in ATE **Connect to Life Science,** p. 99 in ATE **Apply,** p. 100 **Eye on the Environment:** Sitting on Your Trash, p. 122	**Section Review,** p. 100 **Quiz,** p. 100 in ATE **Alternative Assessment,** p. 100 in ATE
Directed Reading Worksheet, Section 2 **Transparency 115,** Porous Rocks Are Reservoirs for Fossil Fuels **Transparency 116,** Formation of Coal **Math Skills for Science Worksheet,** Parts of 100: Calculating Percentages **Reinforcement Worksheet,** If It's a Fossil, How Is It a Fuel?	**Chemistry Connection,** p. 102 **Multicultural Connection,** p. 102 in ATE **Cross-Disciplinary Focus,** p. 102 in ATE **Real-World Connection,** p. 104 in ATE **Math and More,** Percent Carbon, p. 105 in ATE **Eureka!** Oil Rush! p. 123	**Section Review,** p. 104 **Homework,** p. 105 in ATE **Section Review,** p. 107 **Quiz,** p. 107 in ATE **Alternative Assessment,** p. 107 in ATE
Directed Reading Worksheet, Section 3 **Math Skills for Science Worksheet,** Radioactive Decay and the Half-life **Transparency 117,** Generating Energy with Fission **Critical Thinking Worksheet,** Nature's Gold	**Cross-Disciplinary Focus,** p. 109 in ATE **Biology Connection,** p. 110 **Connect to Astronomy,** p. 110 in ATE **Real-World Connection,** p. 111 in ATE **Cross-Disciplinary Focus,** p. 112 in ATE **Connect to Physical Science,** p. 113 in ATE **Multicultural Connection,** p. 113 in ATE **MathBreak,** Miles per Acre, p. 114 **Multicultural Connection,** p. 114 in ATE	**Section Review,** p. 112 **Self-Check,** p. 113 **Homework,** p. 113 in ATE **Section Review,** p. 115 **Quiz,** p. 115 in ATE **Alternative Assessment,** p. 115 in ATE

END-OF-CHAPTER REVIEW AND ASSESSMENT

Chapter Review in Study Guide
Vocabulary and Notes in Study Guide
Chapter Tests with Performance-Based Assessment, Chapter 5 Test
Chapter Tests with Performance-Based Assessment, Performance-Based Assessment 5
Concept Mapping Transparency 5

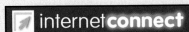 internet **connect**

 go.hrw.com **Holt, Rinehart and Winston On-line Resources**
go.hrw.com

For worksheets and other teaching aids related to this chapter, visit the HRW Web site and type in the keyword: **HSTENR**

 SC**LINKS** NSTA **National Science Teachers Association**
www.scilinks.org

Encourage students to use the *sci*LINKS numbers listed in the internet connect boxes to access information and resources on the **NSTA** Web Site.

Chapter Resources & Worksheets

Visual Resources

TEACHING TRANSPARENCIES

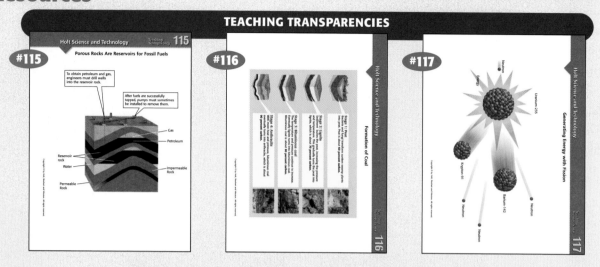

#115 Porous Rocks Are Reservoirs for Fossil Fuels

#116 Formation of Coal

#117 Generating Energy with Fission

TEACHING TRANSPARENCIES

#77 Practicing Conservation

Reduce · Reuse · Recycle

CONCEPT MAPPING TRANSPARENCY

#5

Holt Science and Technology   Concept Mapping Transparency 5

Energy Resources

Use the following terms to complete the concept map below:
biomass, coal, natural gas, nonrenewable, natural resources, renewable, fossil fuels, wind energy

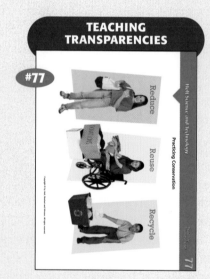
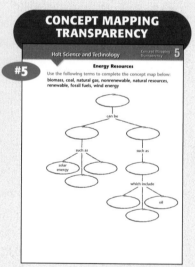

Meeting Individual Needs

DIRECTED READING

#5 DIRECTED READING WORKSHEET

Energy Resources

Chapter Introduction
As you begin this chapter, answer the following.
1. Read the title of the chapter. List three things that you already know about this subject.

2. Write two questions about this subject that you would like answered by the time you finish this chapter.

3. How does the title of the Start-Up Activity relate to the subject of the chapter?

Section 1: Natural Resources (p. 96)
4. Which of the following items does the atmosphere contribute to our "life support system" on Earth? (Circle all that apply.)
   a. shelter   d. rain
   b. food   e. air
   c. warmth   f. nutrients
5. Humans _____ their immediate surroundings by using natural resources. (change or adapt to)
6. Figure 1 shows products made from _____

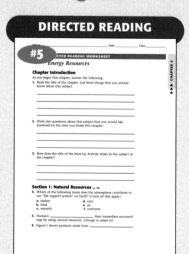

REINFORCEMENT & VOCABULARY REVIEW

#5 REINFORCEMENT WORKSHEET

What Are My Resources?

Complete this worksheet after you finish reading Chapter 5, Section 1.
Something that comes from the Earth and that people use is known as a **natural resource**. There are lots of natural resources on the Earth and they are broken up into two types—**renewable** and **nonrenewable**. You might be wondering what the difference is between these two. It is actually quite simple and has to do with how long it takes to replace the resource. Renewable resources, such as trees, are replaced in a relatively short time after they are used. But a nonrenewable resource, like coal, can take thousands or millions of years to replace. Because it takes such a long time to replace nonrenewable resources, whatever amount exists on Earth right now is limited. This means that once the resource is gone we can't possibly wait until more is made because it just takes too long.

Take a look at the pictures below and label each item with an *R* if it is renewable, or with an *N* if it is nonrenewable. Write your answer on the line provided.

VOCABULARY REVIEW WORKSHEET

#5 VOCABULARY REVIEW WORKSHEET

Energetic Anagram

After finishing Chapter 5, give this puzzle a try!
Use the clues in the first column to unscramble the words in the second column. When you are done, use the letters marked with the numbers 1 through 4 to fill in the statement at the bottom of the page by unscrambling each number grouping to create a new word.

1. A material from the Earth that is used by people: TANLUAR CSOBUREE

2. Examples are trees and fresh water: BAREEENLW   OEECSRUR

3. Examples are oil and natural gas: BELWENNEARON   SEORCUER

4. An example of this is using old newspapers to make the paper for new newspapers: YIERCGLCN

5. A natural resource that people use to make energy: GREENY   EERUOCSR

6. These were once living organisms but their remains were buried: ISLSFO   UFEL

7. Gasoline, jet fuel, and diesel fuel are all made from this: TPEUROMEL

8. Used to generate electricity or for heating: UATLNAR   AGS

9. The only fossil fuel that is a rock: ALOC

10. The method used to extract shallow deposits of coal from the Earth: STPRI   INMGIN

11. This is rain or snow with a high amount of sulfuric acid in it: ADCI   REIPNTTECIAPO

SCIENCE PUZZLERS, TWISTERS & TEASERS

#5 SCIENCE PUZZLERS, TWISTERS & TEASERS

Energy Resources

Fractured Frames
1. Each frame represents a word from the chapter. If you read it in just the right way, what word or phrase does each puzzle represent?

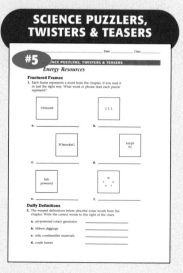

Daffy Definitions
2. The warped definitions below describe some words from the chapter. Write the correct words to the right of the clues.
   a. air-powered rotary generator
   b. ribbon diggings
   c. relic combustible materials
   d. crude fumes

Review & Assessment

STUDY GUIDE

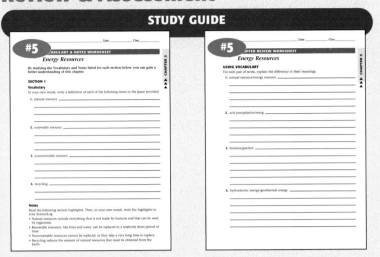

CHAPTER TESTS WITH PERFORMANCE-BASED ASSESSMENT

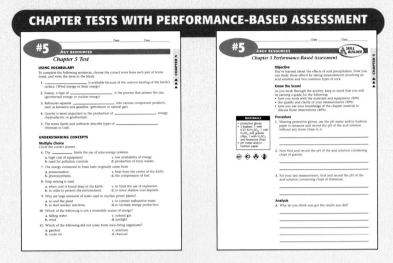

Lab Worksheets

LONG-TERM PROJECTS & RESEARCH IDEAS

DATASHEETS FOR LABBOOK

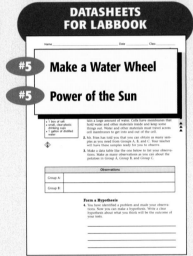

#5 Make a Water Wheel

#5 Power of the Sun

Applications & Extensions

CRITICAL THINKING & PROBLEM SOLVING

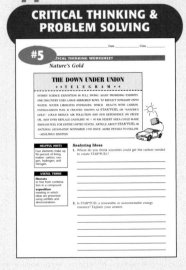

MULTICULTURAL CONNECTIONS

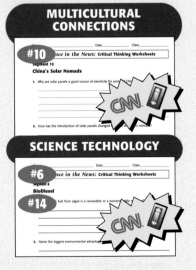

#10 Science in the News: Critical Thinking Worksheets

Segment 10
China's Solar Nomads

SCIENCE TECHNOLOGY

#6 Science in the News: Critical Thinking Worksheets

Segment 6
BioDiesel

#14

SCIENTISTS IN ACTION

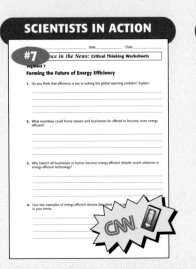

#7 Science in the News: Critical Thinking Worksheets

Segment 7
Forming the Future of Energy Efficiency

INTERACTIVE EXPLORATIONS

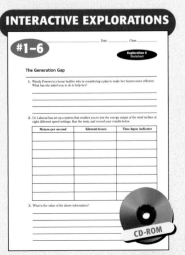

#1–6

The Generation Gap

Chapter Background

SECTION 1

Natural Resources

▶ Interconnected Resources

The Earth's resources are intricately linked. For example, clearing trees affects the water quality downstream from a forest. We should pay attention not only to the rate at which we use natural resources but also to how our use  affects other resources. The use of fossil fuels offers another example; fossil-fuel combustion adds enormous quantities of carbon dioxide, a greenhouse gas, to the atmosphere. Many scientists think that rising carbon dioxide levels are linked to the rising temperatures the Earth has experienced in recent years.

▶ The Three Rs

Over the past three decades, environmentalists have encouraged consumers to consider the three Rs—reduce, recycle, and reuse—to conserve Earth's natural resources.

- When buying new items, consumers who want to save money and conserve resources should buy in bulk or purchase products with minimal packaging. Washing and reusing plastic  bags is another way to conserve petroleum-based natural resources. Use of cloth rags and napkins instead of paper towels and paper napkins is one way to reduce paper consumption. Glass and ceramic plates and cups and metal cutlery are more environmentally responsible than paper plates and plastic cups and utensils. Donating used items to charities and friends and family members is one way to recycle clothes, shoes, appliances, books, toys, and other such items.

SECTION 2

Fossil Fuels

▶ Light Up Your Life

Before automobiles were invented and before electricity was discovered, one of the primary functions of fossil fuels was to provide light. Kerosene lamps became popular in the United States after the first oil well was drilled in 1859 in Pennsylvania. At the same time, use of coal gas and natural gas in lamps was increasing. Coal gas had been used in lamps as early as 1784. By the early 1800s, most cities in the United States and Europe had coal-gas street lights. Electric lamps did not replace the gaslights until the early 1900s.

IS THAT A FACT!

- ➤ Ancient fossil reefs buried underground make excellent oil and gas reservoirs because the reefs are very porous. The productivity of the oil fields in Alberta, Canada, is due to the presence of Devonian reefs that are 408 million to 360 million years old.

▶ "Rigs-to-Reefs"

Environmentalists usually consider offshore oil rigs to be detrimental to marine ecosystems because they disturb animal life and pose a risk of oil spills. Since 1979, however, obsolete oil rigs  have become, many people believe, a welcome addition to these areas. In 1979, the Rigs-to-Reefs program was initiated when a rig was moved from offshore Louisiana to a designated site off Florida to become an artificial reef. During the next 20 years, more than 500 platforms were relocated for the same purpose.

- Within 6 months of placement in a suitable marine area, a platform is covered with invertebrates and plants. These organisms attract other invertebrates and fish, forming the basis of a complex food chain. The open framework of the rig allows water to circulate and fish to swim freely through the structure. Marine animals and commercial and recreational anglers benefit from the artificial reefs. Some people remain concerned, however, about residual pollution from the submerged rigs.

Alternative Resources

▶ Chernobyl

The radioactive fallout from the 1986 Chernobyl nuclear accident affected people, livestock, and crops.

- Although only 31 people died from direct exposure, about 600,000 people were "significantly exposed" to the fallout. At least 50,000 people received 50 rads of radiation. About 10,000 people were exposed to at least 100 rads and suffered from radiation sickness. For comparison, a person undergoing a chest X ray is exposed to a maximum of 1 rad.

- Livestock also suffered from the fallout. At least 86,000 head of cattle were evacuated from the area immediately after the accident. The sale of milk, meat, and many fruits and vegetables was banned in 1986 and 1987 in cities near Chernobyl. Many countries across Europe lost crops and other kinds of vegetation due to radioactive contamination.

▶ Concentrating Solar Power Systems

Concentrating solar power systems harness solar energy by focusing reflected sunlight onto a receiver. The receiver absorbs the light and converts it into thermal energy, which is then used to generate electricity.

- **Solar trough systems** consist of parabolic, mirrored troughs that focus sunlight onto oil-filled tubes at the troughs' focal points. The sunlight heats the oil, which then heats water. Steam from the heated water turns turbines in a generator to produce electricity.

- **Solar power towers** use thousands of mirrors to reflect sunlight onto a receiver that is mounted on a tall tower. The receiver contains salt that stores the solar energy as thermal energy. The salt heats water, creating steam that turns turbines to produce electricity.

- **Solar dish systems** use circular mirrors arranged into the shape of a dish to concentrate solar energy onto a receiver. The receiver transfers the energy to an engine that generates electricity.

▶ Geothermal Energy

Geothermal energy is tapped from places on Earth that are heated by their proximity to volcanic activity. It is currently used in Japan, parts of Russia, Iceland, Italy, New Zealand, and on the western coast of the United States.

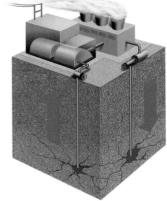

- Iceland is the world leader in using geothermal energy for space heating. Over 85 percent of Icelanders use geothermal energy to warm their homes! The cost is only about one-third of what it would cost if they burned oil to power electric heaters. Industries in Iceland use geothermal energy because it is inexpensive, widely available, and very reliable.

IS THAT A FACT!

➥ Traditional fuels, such as wood and animal dung, are used to meet one-quarter of India's energy needs.

For background information about teaching strategies and issues, refer to the **Professional Reference for Teachers.**

 Pre-Reading Questions

Students may not know the answers to these questions before reading the chapter, so accept any reasonable response.

Suggested Answers

1. Answers will vary. Sample answer: coal, petroleum, iron ore, diamonds

2. fossil fuels

3. A solar cell is an individual cell that converts light into electrical energy, while a solar panel is a panel composed of many solar cells wired together.

CHAPTER

5

Energy Resources

Sections

 Pre-Reading Questions

1. List four nonrenewable resources.

2. On which energy resources do humans currently depend the most?

3. What is the difference between a solar cell and a solar panel?

LIVING INSIDE YOUR TRASH?

Would you believe that this house is made from empty soda cans and old tires? Well, it is! Not only does this house use recycled materials, but it also saves Earth's energy resources. This house gets all its energy from the sun and uses rainwater for household activities. In this chapter, you will learn about what Earth's energy resources are and how we can conserve them.

Architect Mike Reynolds in front of The Castle in Taos, New Mexico

 internet connect

 HRW On-line Resources

go.hrw.com

For worksheets and other teaching aids, visit the HRW Web site and type in the keyword: **HSTENR**

 SciLINKS
NSTA

www.scilinks.com

Use the *sci*LINKS numbers at the end of each chapter for additional resources on the **NSTA** Web site.

 Smithsonian Institution

www.si.edu/hrw

Visit the Smithsonian Institution Web site for related on-line resources.

CNNfyi.com

www.cnnfyi.com

Visit the CNN Web site for current events coverage and classroom resources.

WHAT IS THE SUN'S FAVORITE COLOR?

Are some colors better than others at absorbing the sun's energy? If so, how might this relate to collecting solar energy? Try the following activity to answer these questions.

Procedure

1. Obtain **at least five balloons** that are the same size and shape. One of the balloons should be white, and one should be black.

2. Place **one large ice cube or several small cubes** in each balloon. Each balloon should contain the same amount of ice.

3. Line the balloons up on a flat, uniformly colored surface that receives direct sunlight. Make sure that all the balloons receive the same amount of sunlight and that the openings in the balloons are not facing directly toward the sun.

4. Keep track of how much time it takes for the ice to melt completely in each of the balloons. You can tell how much ice has melted in each balloon by pinching the balloon's opening and then gently squeezing the balloon.

Analysis

5. In which balloon did the ice melt first? Why?

6. What color would you paint a device used to collect solar energy?

97

WHAT IS THE SUN'S FAVORITE COLOR?

MATERIALS
FOR EACH GROUP:
• 5 different-colored, round balloons
• scissors
• ice cubes
• watch or clock with a second hand

Teacher's Notes

Balloons must be identical except for color. Also, note that one balloon in each group must be black and one must be white.

A single, large ice cube or several small cubes are enough for each balloon. Stress to students that they should not use too much ice or it will take too long to get results. You might have students mark their balloons with permanent markers before they add the ice so that each group is able to identify its balloons.

Answers to START-UP Activity

5. The ice in the black balloon melted first; the darker an object is, the more light energy it will absorb.

6. black

Natural Resources

In this section, students learn the difference between renewable and nonrenewable resources. They also learn why conservation and recycling are important.

Bellringer

Display the following items:

a plastic sandwich bag, a piece of paper, a pencil, a glass of water, 1 qt of motor oil, an empty soda can, a wooden match, a salt shaker, and some aquarium charcoal

Challenge students to determine what all of these items have in common. Lead students to conclude that all these items have their origin in natural resources. Have students help you make a list of which resources are renewable and which are nonrenewable.
Sheltered English

1 Motivate

DISCUSSION

Kinds of Energy Have students brainstorm to form a list of different kinds of energy, including light energy, chemical energy, potential energy, kinetic energy, and thermal energy. Review the meaning of each term, if necessary. Now is also a good time to review the law of conservation of energy, which states that energy is never created or destroyed—it can only be changed from one form to another.

**Directed Reading
Worksheet** Section 1

Terms to Learn

natural resource
renewable resource
nonrenewable resource
recycling

What You'll Do

◆ Determine how humans use natural resources.
◆ Contrast renewable resources with nonrenewable resources.
◆ Explain how humans can conserve natural resources.

Natural Resources

Think of the Earth as a giant life-support system for all of humanity. The Earth's atmosphere, waters, and solid crust provide almost everything we need to survive. The atmosphere provides the air we need to breathe, maintains air temperatures, and produces rain. The oceans and other waters of the Earth provide food and needed fluids. The solid part of the Earth provides nutrients and minerals.

Interactions between the Earth's systems can cause changes in the Earth's environments. Organisms must adapt to these changes if they are to survive. Humans have found ways to survive by using natural resources to change their immediate surroundings. A **natural resource** is any natural substance, organism, or energy form that living things use. Few of the Earth's natural resources are used in their unaltered state. Most resources are made into products that make people's lives more comfortable and convenient, as shown in **Figure 1**.

Figure 1 *Lumber, gasoline, and electricity are all products that come from natural resources.*

This pile of lumber is made of wood, which comes from trees.

The gasoline in this can is made from oil pumped from the Earth's crust.

Electricity generated by these wind turbines ultimately comes from the sun's energy.

98

CONNECT TO ENVIRONMENTAL SCIENCE

There are numerous opportunities for students to become involved in volunteer efforts to conserve and recycle natural resources. As a class, find out what opportunities there are in your area. Students may be interested in a beach or river cleanup, tree planting, or a resource conservation public outreach program. Suggest that students volunteer two weekends as they study this chapter. At the end of the chapter, students can share their volunteer experiences with the class and present opportunities for their classmates to become involved.

Renewable Resources

Some natural resources are renewable. A **renewable resource** is a natural resource that can be used and replaced over a relatively short time. **Figure 2** shows two examples of renewable resources. Although many resources are renewable, humans often use them more quickly than they can be replaced. Trees, for example, are renewable, but humans are currently cutting trees down more quickly than other trees can grow to replace them.

Figure 2 *Fresh water and trees are just a few of the renewable resources available on Earth.*

Nonrenewable Resources

Not all of Earth's natural resources are renewable. A **nonrenewable resource** is a natural resource that cannot be replaced or that can be replaced only over thousands or millions of years. Examples of nonrenewable resources are shown in **Figure 3.** The amounts of nonrenewable resources on Earth are fixed with respect to their availability for human use. Once nonrenewable resources are used up, they are no longer available. Oil and natural gas, for example, exist in limited quantities. When these resources become scarce, humans will have to find other resources to replace them.

Figure 3 *Nonrenewable resources, such as coal and natural gas, can be replaced only over thousands or millions of years once they are used up.*

CONNECT TO
LIFE SCIENCE

Round out your discussion of nonrenewable resources by introducing students to the three Rs of conservation. Use Teaching Transparency 77 as a guide to help students list as many examples as possible of reducing, reusing, and recycling.

Teaching Transparency 77
"Practicing Conservation"

LINK TO LIFE SCIENCE

2) Teach

REAL-WORLD CONNECTION

Aluminum is refined from the ore bauxite, which is deposited in a thin layer at the Earth's surface. Worldwide, bauxite strip mines cover more of the Earth's surface than any other type of metal ore mine. Aluminum production uses so much electrical power that the metal has been referred to as "congealed electricity." To produce six aluminum cans, it takes the energy equivalent of 1 L of gasoline. For this reason, aluminum smelters are located close to sources of cheap, reliable power, such as the hydroelectric dams in the Pacific Northwest, Quebec, and the Amazon. When the environmental damage caused by producing new aluminum is considered, the importance of recycling becomes clear. Recycling one aluminum can saves enough energy to run a television set for 4 hours! Currently, the United States obtains about 20 percent of its aluminum from recycling. Have students write a persuasive letter explaining why that percentage should increase.

MEETING INDIVIDUAL NEEDS

Learners Having Difficulty Have students help you define renewable and nonrenewable resources. Provide students with two pieces of poster board, scissors, glue, and magazines. Have them attach pictures of products made from renewable resources to one poster board and pictures of products made from nonrenewable resources to the other. Beside each type of resource, have them describe where it is from, how it is obtained, the environmental effects of extracting and using the resource, and have them suggest ways it can be conserved or recycled.

3 Close

Quiz

1. Explain why sunlight is our most valuable natural resource. (Light from the sun is an energy source that plants use to create food and that we can harness to do work. Students may also note that the energy in fossil fuels has its origin in sunlight.)

2. Explain the difference between conserving a resource and recycling it. (Conserving a resource means using it sparingly and not wasting it. Recycling refers to the reuse of natural resources to make new products.)

ALTERNATIVE ASSESSMENT

Concept Mapping Have students use the following terms to construct a concept map:

use, disposal, recycling, nonrenewable resource, reuse, production, renewable resource

Reinforcement Worksheet
"What Are My Resources?"

TOPIC: Natural Resources
GO TO: www.scilinks.org
*sci*LINKS NUMBER: HSTE105

Renewable or Nonrenewable?

Find five products in your home that were made from natural resources. List the resource or resources from which each product was made. Label each resource as renewable or nonrenewable.

Are the products made from mostly renewable or nonrenewable resources? Are those renewable resources plentiful on Earth? Do humans use those renewable resources more quickly than the resources can be replaced? What can you do to help conserve nonrenewable resources and renewable resources that are becoming more scarce?

Figure 4 *You can recycle many household items to help conserve natural resources.*

internet**connect**

SCILINKS.
NSTA

TOPIC: Natural Resources
GO TO: www.scilinks.org
*sci*LINKS NUMBER: HSTE105

Conserving Natural Resources

Whether the natural resources we use are renewable or nonrenewable, we should be careful how we use them. To conserve natural resources, we should try to use them only when necessary. For example, leaving the faucet running while brushing your teeth wastes clean water. Turning the faucet on only to rinse your brush saves a lot of water that you or others need for other uses.

Another way to conserve natural resources is to recycle, as shown in **Figure 4.** **Recycling** is the process by which used or discarded materials are treated for reuse. Recycling allows manufacturers to reuse natural resources when making new products. This in turn reduces the amount of natural resources that must be obtained from the Earth. For example, recycling aluminum cans reduces the amount of aluminum that must be mined from the Earth's crust to make new cans.

SECTION REVIEW

1. How do humans use most natural resources?

2. What is the difference between renewable and nonrenewable resources?

3. Name two ways to conserve natural resources.

4. **Applying Concepts** List three renewable resources not mentioned in this section.

100

▼ Answers to Section Review

1. Humans consume most natural resources by using products made from the resources.

2. Renewable resources can be replaced over a relatively short time, while nonrenewable resources cannot be replaced or can be replaced only over thousands or millions of years.

3. We can conserve natural resources by using them only when necessary and by recycling them.

4. Answers will vary.

Fossil Fuels

Terms to Learn

energy resource   coal
fossil fuel   strip mining
petroleum   acid precipitation
natural gas   smog

What You'll Do

◆ Classify the different forms of fossil fuels.
◆ Explain how fossil fuels are obtained.
◆ Identify problems with fossil fuels.
◆ List ways to deal with fossil-fuel problems.

Energy resources are natural resources that humans use to produce energy. There are many types of renewable and nonrenewable energy resources, and all of the energy released from these resources ultimately comes from the sun. The energy resources on which humans currently depend the most are fossil fuels. **Fossil fuels** are nonrenewable energy resources that form in the Earth's crust over millions of years from the buried remains of once-living organisms. Energy is released from fossil fuels when they are burned. There are many types of fossil fuels, which exist as liquids, gases, and solids, and humans use a variety of methods to obtain and process them. These methods depend on the type of fossil fuel, where the fossil fuel is located, and how the fossil fuel formed. Unfortunately, the methods of obtaining and using fossil fuels can have negative effects on the environment. Read on to learn about fossil fuels and the role they play in our lives.

Liquid Fossil Fuels—Petroleum

Petroleum, or crude oil, is an oily mixture of flammable organic compounds from which liquid fossil fuels and other products, such as asphalt, are separated. Petroleum is separated into several types of fossil fuels and other products in refineries, such as the one shown in **Figure 5.** Among the types of fossil fuels separated from petroleum are gasoline, jet fuel, kerosene, diesel fuel, and fuel oil.

Figure 5 *Fossil fuels and other products are separated from petroleum in a process called fractionation. In this process, petroleum is gradually heated in a tower so that different components boil and vaporize at different temperatures.*

101

Focus

Fossil Fuels

In this section, students will learn how fossil fuels, such as petroleum, natural gas, and coal, form and where deposits of these fuels are found in the United States. Students will also learn about some of the ways we obtain fossil fuels and about the environmental problems associated with obtaining and using fossil fuels.

🔔 Bellringer

Pose this question on the board or an overhead projector:

What does the term *fossil fuels* imply about the source of these fuels? (The term *fossil fuels* implies that these fuels are derived from the remains of ancient life.)

1 Motivate

DISCUSSION

Fossil Fuel Use Lead students in a discussion about why fossil fuels, such as coal and gasoline, are so widely used as energy resources. (Answers may include their cost, availability, and ease of use.)

Challenge students to think about what qualities a good fuel should have. (It should be abundant, affordable, easy to obtain, have a high ratio of energy to weight, be easy to transport, and produce little waste.)

Give students 10 minutes to write about how the disadvantages of fossil fuel use affect them.

Science BIOopers

On January 10, 1901, oil from the famous Spindletop well near Beaumont, Texas, began to flow. In fact, the crude oil spewed higher than 90 m into the air! Caught off guard by the tremendous volume of petroleum, 100,000 barrels per day, drillers took 9 days to cap the well.

IS THAT A FACT!

During the early 1900s, petroleum was so plentiful in the United States that it sold for about 3¢ per barrel. At about the same time, water—a renewable resource—sold for about $6 per barrel!

Directed Reading Worksheet Section 2

2) Teach

BRAIN FOOD

One characteristic all fossil fuels share is that they are formed from the remains of organisms that lived long ago. Over millions of years, these remains are buried by sediment and rock. Rising pressure and temperature cause slow chemical changes that result in coal, petroleum, or natural gas. When hydrocarbons burn, or combine rapidly with oxygen, they release energy. The energy that originally formed the carbon-hydrogen bonds was captured by plants and incorporated in their tissues by photosynthesis.

DISCUSSION

Tell students that combustion of natural gas creates very little soot, carbon monoxide, or nitrogen oxides—it burns "cleaner" than other fossil fuels. Then ask students why industrialized societies have developed a reliance on solid and liquid fossil fuels rather than on gaseous fossil fuels. Provide a hint by asking whether it is easier to design a leak-proof container for 1 L of milk or 1 L of oxygen.

Multicultural CONNECTION

Natural gas seeps were first discovered in ancient Persia (now Iran) between 6000 and 2000 B.C. Records from China indicate use of natural gas by 900 B.C. The Chinese drilled the first known natural gas well using bamboo poles and primitive drill bits. The well was 140 m deep. In Europe, natural gas was first discovered in England in 1659.

Chemistry CONNECTION

Petroleum and natural gas are both made of compounds called hydrocarbons. A *hydrocarbon* is an organic compound containing only carbon and hydrogen.

Gaseous Fossil Fuels—Natural Gas

Gaseous fossil fuels are classified as **natural gas.** Most natural gas is used for heating and for generating electricity. The stove in your kitchen may be powered by natural gas. Many motor vehicles, such as the van in **Figure 6,** are fueled by liquefied natural gas. Vehicles like these produce less air pollution than vehicles powered by gasoline.

Methane is the main component of natural gas. But other natural-gas components, such as butane and propane, can be separated and used by humans. Butane is often used as fuel for camp stoves. Propane is used as a heating fuel and as a cooking fuel, especially for outdoor grills.

Figure 6 *Vehicles powered by liquefied natural gas are becoming more common.*

Figure 7 *This coal is being gathered so that it may be burned in the power plant shown in the background.*

Solid Fossil Fuels—Coal

The solid fossil fuel that humans use most is coal. **Coal** is a solid fossil fuel formed underground from buried, decomposed plant material. Coal, the only fossil fuel that is a rock, was once the leading source of energy in the United States. People burned coal for heating and transportation. Many trains in the 1800s and early 1900s were powered by coal-burning steam locomotives.

People began to use coal less because burning coal often produces large amounts of air pollution and because better energy resources were discovered. Coal is no longer used much as a fuel for heating or transportation in the United States. However, many power plants, like the one shown in **Figure 7,** burn coal to produce electricity.

CROSS-DISCIPLINARY FOCUS

Writing **Language Arts** Petroleum is more than a fuel source; it is refined to make plastics and other petrochemical products we use every day. Petrochemicals are used to create medicines, inks, solvents, clothing, fertilizers, and many other products. Have students make a list of 20 products made from petroleum. Then have students write a short story that is based in a world with no petroleum products. They can set their stories in the nineteenth century, when petrochemical products did not exist, or in a world of the future when petroleum reserves have been exhausted.

How Do Fossil Fuels Form?

All fossil fuels form from the buried remains of ancient organisms. But different types of fossil fuels form in different ways and from different types of organisms. Petroleum and natural gas form mainly from the remains of microscopic sea life. When these organisms die, their remains settle on the ocean floor, where they decay and become part of the ocean sediment. Over time, the sediment slowly becomes rock, trapping the decayed remains. Through physical and chemical changes over millions of years, the remains become petroleum and gas. Gradually, more rocks form above the rocks that contain the fossil fuels. Under the pressure of overlying rocks and sediments, the fossil fuels are squeezed out of their source rocks and into permeable rocks. As shown in **Figure 8,** these permeable rocks become reservoirs for petroleum and natural gas. The formation of petroleum and natural gas is an ongoing process. Part of the remains of today's sea life will probably become petroleum and natural gas millions of years from now.

Rock Sponge

1. Place samples of **sandstone, limestone,** and **shale** in separate **Petri dishes.**

2. Place 5 drops of light **machine oil** on each rock sample.

3. Observe and record the time required for the oil to be absorbed by each of the rock samples.

4. Which rock sample absorbed the oil fastest? Why?

5. Based on your findings, describe a property that allows for easy removal of fossil fuels from reservoir rock.

To obtain petroleum and gas, engineers must drill wells into the reservoir rock.

After fuels are successfully tapped, pumps must sometimes be installed to remove them.

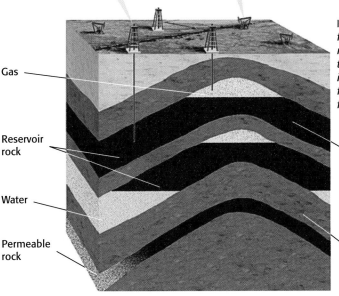

Figure 8 *Petroleum and gas rise from source rock into reservoir rock. Sometimes the fuels are trapped by overlying rock that is impermeable. Rocks that are folded upward are excellent fossil-fuel traps.*

Gas

Reservoir rock

Water

Permeable rock

Petroleum

Impermeable rock

103

internetconnect

SC*LINKS*
NSTA

TOPIC: Fossil Fuels
GO TO: www.scilinks.org
*sci*LINKS NUMBER: HSTE120

Teaching Transparency 115 "Porous Rocks Are Reservoirs for Fossil Fuels"

QuickLab

MATERIALS

FOR EACH STUDENT:
- samples of sandstone, limestone, and shale
- Petri dishes
- light machine oil
- eyedropper

Answers to QuickLab

4. Answers will vary. The rock sample with the highest percentage of interconnected pore space should soak up the oil the fastest. Most reservoir rock is limestone that has many connected pores.

5. Oil and gas move easily through reservoir rock that has a high percentage of interconnected pore space. This allows for easy removal of liquid fossil fuels.

DEMONSTRATION

MATERIALS

- clean, empty glass jar
- 100 mL each of rubbing alcohol, water, and vegetable oil
- red and blue food coloring

Simulating Reservoirs Petroleum does not fill all the available pore space in a subsurface reservoir. Natural gas often accumulates above an oil deposit, as shown in **Figure 8.** When it is located on top of an oil deposit, natural gas can be obtained through shallow drilling.

Tell students that the alcohol represents natural gas and that the cooking oil represents petroleum. Add a few drops of blue color to the water and a few drops of red color to the alcohol. Add the liquids to the jars in different sequences, and ask if this different treatment affects the final results. (no)
Sheltered English

Section 2 • Fossil Fuels **103**

2 Teach, continued

USING THE FIGURE

Have students answer the questions below.

- What kinds of organisms play an important role in coal production? (bacteria, plants, and fungi)
- In a given area, which would be older: peat or lignite deposits? (Lignite; peat is an earlier stage of coal formation.)

REAL-WORLD CONNECTION

Have students find out what fuel sources are used to produce electricity in their community. If fossil fuels are used, have students identify the fuel's source and explain how the fuel is transported to the power plant. As an extension, have students create a list of 15 ways electricity consumption in your community can be reduced. Students could create sign-up sheets and see how many people they can find to pledge to follow their 15-step energy conservation plan.

INDEPENDENT PRACTICE

Concept Mapping Have students construct concept maps that show the different methods of extraction used for various fossil fuels. Their maps should include the formation of the fuel, the location, and the extraction system. One type of fuel may have two or more locations and extraction methods. Encourage students to do additional research to complete their maps.

Teaching Transparency 116
"Formation of Coal"

Coal Formation Coal forms differently from petroleum and natural gas. Coal forms underground over millions of years from decayed swamp plants. When swamp plants die, they sink to the bottom of the swamps. This begins the process of coal formation, which is illustrated below. Notice that the percentage of carbon increases with each stage. The higher the carbon content, the cleaner the material burns. However, all grades of coal will pollute the air when burned.

The Process of Coal Formation

Stage 1: Peat
Bacteria and fungi transform sunken swamp plants into peat. Peat is about **60 percent carbon.**

Stage 2: Lignite
Sediment buries the peat, increasing the pressure and temperature. This gradually turns the peat into lignite, which is about **70 percent carbon.**

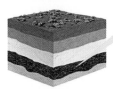

Stage 3: Bituminous coal
The temperature and pressure continue to increase. Eventually lignite turns into bituminous coal. Bituminous coal is about **80 percent carbon.**

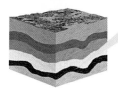

Stage 4: Anthracite
With more heat and pressure, bituminous coal eventually turns into anthracite, which is about **90 percent carbon.**

SECTION REVIEW

1. Name a solid, liquid, and gaseous fossil fuel.
2. What component of coal-forming organic material increases with each step in coal formation?
3. **Comparing Concepts** What is the difference between the organic material from which coal forms and the organic material from which petroleum and natural gas mainly form?

104

▼ **Answers to Section Review**

1. Answers will vary. The solid fossil fuel mentioned in this section is coal. Liquid fossil fuels mentioned in this section include gasoline, jet fuel, kerosene, diesel fuel, and fuel oil (petroleum is an acceptable answer). Gaseous fossil fuels mentioned in this section include methane, butane, and propane (natural gas is an acceptable answer).

2. carbon
3. Coal forms from buried, decayed swamp plants, while oil and natural gas form mainly from buried, decayed sea life.

Where Are Fossil Fuels Found?

Fossil fuels are found in many parts of the world, both on land and beneath the ocean. As shown in **Figure 9,** the United States has large reserves of petroleum, natural gas, and coal. In spite of all our petroleum reserves, we import about one-half of our petroleum and petroleum products from the Middle East, South America, and Africa.

Key
- Petroleum
- Natural gas
- Coal

Figure 9 *Most oil and gas produced in the continental United States comes from California, Louisiana, and Texas.*

How Do Humans Obtain Fossil Fuels?

Humans use different methods to remove fossil fuels from the Earth's crust. These methods depend on the type of fuel being obtained and its location. Petroleum and natural gas are removed from the Earth by drilling wells into rock that contains these resources. Oil wells exist both on land and in the ocean. For offshore drilling, engineers mount drills on platforms that are secured to the ocean floor or float at the ocean's surface. **Figure 10** shows an offshore oil rig.

Coal is obtained either by mining deep beneath the Earth's surface or by strip mining. **Strip mining** is a process in which rock and soil are stripped from the Earth's surface to expose the underlying materials to be mined. Strip mining is used to mine shallow coal deposits. **Figure 11** shows a coal strip mine.

Figure 10 *Large oil rigs, some more than 300 m tall, operate offshore in many places, such as the Gulf of Mexico and the North Sea.*

Figure 11 *Strip miners use explosives to blast away rock and soil and to expose the material to be mined.*

105

Homework

Fuel-Efficient Cars In 1998, the United States consumed nearly 583 billion liters of liquid fossil fuels. More than 460 billion liters of this total was gasoline. After an average car is driven 11,000 mi, it has released its weight in carbon dioxide. Have students compare the fuel efficiency of commercially available cars and calculate the annual carbon dioxide reduction and fuel savings a fuel-efficient car enables.

DEBATE

Drilling in a Wildlife Refuge

The U.S. Fish and Wildlife Service, which administers Alaska's Arctic National Wildlife Refuge, states that its primary mandate is "to protect the wildlife and habitats of this area for the benefit of people now and in the future." The refuge's coastal plain is the calving ground for the Porcupine caribou herd, the most important land-based denning area for the entire Beaufort Sea polar bear population, home for 350 reintroduced musk oxen, and an important habitat for more than 180 bird species. Environmentalists claim that oil drilling on the refuge would bring pollution and disrupt the lives of the animals that use the coastal plain. Oil-industry executives say that they would drill on only 8 percent of the refuge (the 1.5-million-acre coastal plain); that oil revenues would benefit the state and federal governments, that more than 250,000 jobs would be created, and that importing foreign oil is too expensive.

Have students learn more about the issue and debate whether or not oil drilling should be allowed in Arctic National Wildlife Refuge.

internet**connect**

SCI**LINKS**
NSTA

TOPIC: Nonrenewable Resources
GO TO: www.scilinks.org
*sci***LINKS NUMBER:** HSTE115

BRAIN FOOD

Although some countries have reduced their use of coal, the known coal reserves will last no more than 250 years at the present rates of coal consumption.

Problems with Fossil Fuels

Although fossil fuels provide energy for our technological world, the methods of obtaining and using them can have negative consequences. For example, when coal is burned, sulfur dioxide is released. Sulfur dioxide combines with moisture in the air to produce sulfuric acid, which is one of the acids in acid precipitation. **Acid precipitation** is rain or snow that has a high acid content due to air pollutants. Acid precipitation negatively affects wildlife, plants, buildings, and statues, as shown in **Figure 12.**

Figure 12 Acid precipitation can dissolve parts of statues.

Coal Mining The mining of coal can also create environmental problems. Strip mining removes soil, which plants need for growth and some animals need for shelter. If land is not properly repaired afterward, strip mining can destroy wildlife habitats. Coal mines that are deep underground, such as the one shown in **Figure 13,** can be hazardous to the men and women working in them. Coal mining can also lower local water tables, pollute water supplies, and cause the overlying earth to collapse.

Petroleum Problems Obtaining petroleum can also cause environmental problems. In 1989, the supertanker *Exxon Valdez* spilled about 260,000 barrels of crude oil into the water when it ran aground off the coast of Alaska. The oil killed millions of animals and damaged the local fishing industry.

Figure 13 Coal dust can damage the human respiratory system. And because coal dust is flammable, it increases the danger of fire and explosion in coal mines.

106

SCIENTISTS AT ODDS

One of the products of burning fossil fuels is carbon dioxide. Scientists recognize carbon dioxide as a greenhouse gas—a gas that traps thermal energy and increases the temperature of the Earth's atmosphere. Most scientists agree that both carbon dioxide levels and global temperatures are increasing.

However, they have different opinions about the many possible ways rising carbon dioxide levels may be affecting the Earth's climate. Have students find newspaper and magazine articles about this issue and write a balanced essay presenting both sides of the debate.

Smog Burning petroleum products causes a big environmental problem called smog. **Smog** is a photochemical fog produced by the reaction of sunlight and air pollutants. Smog is particularly serious in places such as Denver and Los Angeles. In these cities, the sun shines most of the time, there are millions of automobiles, and surrounding mountains prevent the wind from blowing pollutants away. Smog levels in some cities, including Denver and Los Angeles, have begun to decrease in recent years.

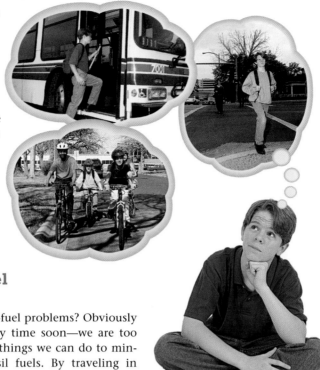

Dealing with Fossil-Fuel Problems

So what can be done to solve fossil-fuel problems? Obviously we can't stop using fossil fuels any time soon—we are too dependent on them. But there are things we can do to minimize the negative effects of fossil fuels. By traveling in automobiles only when absolutely necessary, people can cut down on car exhaust in the air. Carpooling, riding a bike, walking, and using mass-transit systems also help by reducing the number of cars on the road. These measures help reduce the negative effects of using fossil fuels, but they do not eliminate the problems. Only by using certain alternative energy resources, which you will learn about in the next section, can we eliminate them.

Figure 14 *Using mass transit, walking, or riding your bike can help reduce air pollution due to burning fossil fuels.*

SECTION REVIEW

1. Name a state with petroleum, natural-gas, and coal reserves.

2. How do we obtain petroleum and natural gas? How do we obtain coal?

3. Name three problems with fossil fuels. Name three ways to minimize the negative effects of fossil fuels.

4. **Making Inferences** Why does the United States import petroleum from other regions even though the United States has its own petroleum reserves?

internet connect

SCiLINKS
NSTA

TOPIC: Fossil Fuels
GO TO: www.scilinks.org
*sci*LINKS **NUMBER:** HSTE120

107

▼ **Answers to Section Review**

1. Answers will vary. Use **Figure 9** to check students' answers.

2. We obtain petroleum and natural gas by drilling wells into rock formations that contain these resources. We obtain coal by mining.

3. Answers will vary. Problems include oil spills, loss of soil from strip mining coal, and the production of smog due to burning

fossil fuels. Ways of minimizing negative effects include carpooling, riding a bike, walking, using public transportation, and recycling products made from petroleum.

4. America imports petroleum because it is cheaper in other regions and because those regions have enacted fewer environmental regulations on petroleum production and transport.

4) Close

Quiz

1. What is the relationship between petroleum and liquid fossil fuels? (Petroleum is the mixture of compounds from which liquid fossil fuels are separated.)

2. Explain why we use different methods to extract fossil fuels from the Earth's crust. (We use different methods because fossil fuels differ in their location and composition.)

3. When an oil and gas reservoir is drilled, which substance is generally encountered first— oil or natural gas? Why? (natural gas; because it is less dense than oil and it migrates to the top of the reservoir)

4. Coal beds have been found in Antarctica. Explain how this could be. (Answers will vary. Antarctica had a warmer climate in the past, and the warmer temperatures allowed plants to grow. These plants were buried and eventually became coal formations.)

ALTERNATIVE ASSESSMENT

Have each student find out about the fossil fuel resources of a particular continent. Ask them to create a map showing where the petroleum, coal, and natural gas deposits are found in the continent. Students can supplement their maps with information and illustrations showing how these resources are extracted, how they are used, and the environmental problems associated with them.

Reinforcement Worksheet
"If It's a Fossil, How Is It a Fuel?"

Focus

Alternative Resources

In this section, students will learn about some of the alternatives to fossil fuels. The section also includes a discussion of the pros and cons of alternative energy sources.

Bellringer

Show students a picture of a wind farm, a solar energy facility, and a hydroelectric dam. Ask them which, if any, of these alternative energy facilities might be well suited to their community. Explain that the energy resources used in these facilities are just some of the alternatives to fossil fuels available.

1 Motivate

DISCUSSION

Have each student make a list of 10 ways in which he or she uses electricity every day. Compile a master list on the board, and discuss alternatives to each activity. (Letting one's hair dry on its own, for example, is an alternative to using a hair dryer.) Have each student try two of these suggestions at home and report back to class.

Directed Reading Worksheet Section 3

Teaching Transparency 117 "Generating Energy with Fission"

Terms to Learn

nuclear energy
solar energy
wind energy
hydroelectric energy

biomass
gasohol
geothermal energy

What You'll Do

- Describe alternatives to the use of fossil fuels.
- List advantages and disadvantages of using alternative energy resources.

Alternative Resources

The energy needs of industry, transportation, and housing are increasingly met by electricity. However, most electricity is currently produced from fossil fuels, which are nonrenewable and cause pollution when burned. For people to continue their present lifestyles, new sources of energy must become available.

Splitting the Atom

Nuclear energy is an alternative source of energy that comes from the nuclei of atoms. Most often it is produced by a process called *fission*. Fission is a process in which the nuclei of radioactive atoms are split and energy is released, as shown in **Figure 15**. Nuclear power plants use radioactive atoms as fuel. When fission takes place, a large amount of energy is released. The energy is used to produce steam to run electric generators in the power plant.

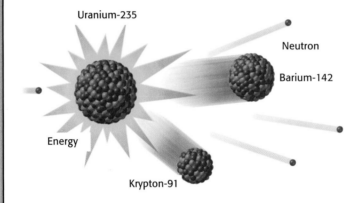

Figure 15 *The process of fission generates a tremendous amount of energy.*

Uranium-235

Neutron

Barium-142

Energy

Krypton-91

Pros and Cons Nuclear power plants provide alternative sources of energy without the problems that come with fossil fuels. So why don't we use nuclear energy instead of fossil fuels? Nuclear power plants produce dangerous wastes. The wastes are unsafe because they are radioactive. Radioactive wastes must be removed from the plant and stored until they lose their radioactivity. But nuclear wastes can remain dangerously radioactive for thousands of years. A safe place must be found to store these wastes so that radiation cannot escape into the environment.

Figure 16 *Areas or objects marked with this symbol should be approached only after taking proper precautions.*

MISCONCEPTION ALERT

Nuclear energy is the energy that exists in the bonds that hold together atomic nuclei. The breaking of these bonds, or nuclear fission, occurs at an uncontrolled rate in an atomic bomb. By contrast, the rate of fission in a nuclear power plant is carefully controlled. As long as nuclear power plants are designed and operated properly, they release safe amounts of radiation and no particulate pollution. It is incredibly efficient: 500 g of uranium-235 has nearly 3 million times the energy-producing potential of the same amount of coal. The greatest problem with nuclear power is locating a safe place to store the waste products generated during fission.

Because nuclear power plants generate a lot of energy, large amounts of water are used in cooling towers, like the ones shown in **Figure 17,** to cool the plants. If a plant's cooling system were to stop working, the plant would overheat, and its reactor could possibly melt. Then a large amount of radiation could escape into the environment, as it did at Chernobyl, Ukraine, in 1986.

Combining Atoms

Another type of nuclear energy is produced by *fusion*. Fusion is the joining of nuclei of small atoms to form larger atoms. This is the same process that is thought to produce energy in the sun.

The main advantage of fusion is that it produces few dangerous wastes. The main disadvantage of fusion is that very high temperatures are required for the reaction to take place. No known material can withstand temperatures that high, so the reaction must occur within a special environment, such as a magnetic field. So far, fusion reactions have been limited to laboratory experiments.

Figure 17 *Cooling towers are one of many safety mechanisms used in nuclear power plants. Their purpose is to prevent the plant from overheating.*

Sitting in the Sun

When sunlight falls on your skin, the warmth you feel is part of solar energy. **Solar energy** is energy from the sun. Every day, the Earth receives more than enough solar energy to meet all of our energy needs. And because the Earth continuously receives solar energy, the energy is a renewable resource.

There are two common ways that we use solar energy. Sunlight can be changed into electricity by the use of solar cells. You may have used a calculator, like the one shown in **Figure 18,** that was powered by solar cells.

Figure 18 *This solar calculator receives all the energy it needs through the four solar cells located above its screen.*

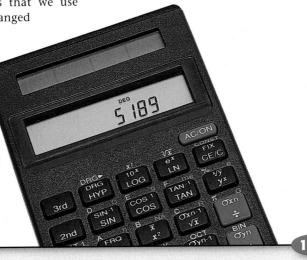

109

READING 📖 STRATEGY

Activity Before they read the information on solar cells and solar collectors, have students deduce whether each of the statements below is true or false. When students have finished reading these two pages, have them correct any wrong answers and rewrite the false statements to make them true. Students can use this activity as a basis for creating an alternative energy misconception fact sheet.

- A solar cell produces only a very small amount of electricity. (true)
- Solar cells are not commonly used because they cause a great deal of pollution. (false)
- The expense of a solar power system does not pay off to consumers over time. (false)
- Solar water heaters are most efficient in the far northern parts of the United States. (false)

MISCONCEPTION ///ALERT\\\

Solar energy is for all practical purposes an inexhaustible source of energy. However, it is important to note that the sun doesn't shine 24 hours a day in most areas nor is it always directly overhead. Using solar energy efficiently depends on the time of day, local weather conditions, the time of year, and an area's latitude. To solve some of these problems, solar power systems use batteries to store the electricity they generate.

Biology
C O N N E C T I O N

Did you know that the energy from petroleum, coal, and natural gas is really a form of stored solar energy? All organisms ultimately get their energy from sunlight and store it in their cells. When ancient organisms died and became trapped in sediment, some of their energy was stored in the fossil fuel that formed in the sediment. So the gasoline that powers today's cars contains energy from sunlight that fell on the Earth millions of years ago!

Figure 20 Perhaps you have seen solar panels used in this manner in your town.

Solar Cells A single solar cell produces only a tiny amount of electricity. For small electronic devices, such as calculators, this is not a problem because enough energy can be obtained with only a few cells. But in order to provide enough electricity for larger objects, such as a house, thousands of cells are needed. Many homes and businesses use solar panels mounted on their roof to provide much of their needed electricity. Solar panels are large panels made up of many solar cells wired together. **Figure 19** shows a building with solar panels.

Figure 19 Although they are expensive to install, solar panels are good investments in the long run.

Counting the Cost Solar cells are reliable and quiet, have no moving parts, and can last for years with little maintenance. They produce no pollution during use, and pollution created by their manufacturing process is very low.

So why doesn't everyone use solar cells? The answer is cost. While solar energy itself is free, solar cells are relatively expensive to make. The cost of a solar-power system could account for one-third of the cost of an entire house. But in remote areas where it is difficult and costly to run electric wires, solar-power systems can be a realistic option. In the United States today, tens of thousands of homes use solar panels to produce electricity. Can you think of other places that you have seen solar panels? Take a look at **Figure 20.**

110

CONNECT TO ASTRONOMY

In space, conserving natural resources is very important. Many of the innovations in alternative energy research began in the space program. For example, NASA scientists must create solar panels that use solar energy as efficiently as possible and design materials that insulate space probes from the temperature extremes of space. Students may be interested in researching the recent NASA development dubbed "aerogel," the lightest solid on Earth. Aerogel, a nearly transparent substance, is only three times as dense as air and has 20 times the insulating power of window glass. Suggest that students share their findings in class.

Solar Heating Another use of solar energy is direct heating through solar collectors. Solar collectors are dark-colored boxes with glass or plastic tops. A common use of solar collectors is heating water, as shown in **Figure 21.** Over 1 million solar water heaters have been installed in the United States. They are especially common in Florida, California, and some southwestern states.

As with solar cells, the problem with solar collectors is cost. But solar collectors quickly pay for themselves—heating water is one of the major uses of electricity in American homes. Also, solar collectors can be used to generate electricity.

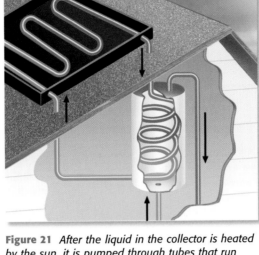

Figure 21 *After the liquid in the collector is heated by the sun, it is pumped through tubes that run through a water heater, causing the temperature of the water to rise.*

Large-Scale Solar Power Experimental solar-power facilities, such as the one shown in **Figure 22,** have shown that it is possible to generate electricity for an entire city. Facilities like this one are designed to use mirrors to focus sunlight onto coated steel pipes filled with synthetic oil. The oil is heated by the sunlight and is then used to heat water. The heated liquid water turns to steam, which is used to drive electric generators.

An alternative design for solar-power facilities is one that uses mirrors to reflect sunlight onto a receiver on a central tower. The receiver captures the sunlight's energy and stores it in tanks of molten salt. The stored energy is then used to create steam, which drives a turbine in an electric generator. *Solar Two,* a solar-power facility designed in this manner, is capable of generating enough energy to power 10,000 homes in southern California.

Turn to page 136 to calculate the power of the sun.

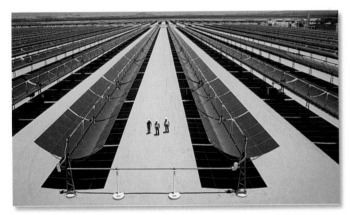

Figure 22 *This solar-power facility in the Mojave Desert has 1,926 sun-tracking mirrors called heliostats.*

111

Power of the Sun

IS THAT A FACT!

In 1980, the production cost of generating electricity from solar energy averaged 60¢ per kilowatt hour. By 1990, costs were reduced to an average of 16¢ per kilowatt hour. Solar energy systems in the United States have a combined capacity of 354 MW!

PG 136

GUIDED PRACTICE

Put the statements below on an overhead transparency in the order shown. Ask students to sequence the statements to explain how the solar power facility in the Mojave Desert works.

- The oil is heated and in turn heats water. (3)
- Computer-guided mirrors collect solar energy. (1)
- The heated water turns into steam, which is used to turn generators that produce electricity. (4)
- Mirrors focus sunlight onto pipes that are filled with oil. (2)

GROUP ACTIVITY

Building a Solar Cooker Solar cookers have been used successfully in many developing countries where deforestation is a problem. There are many different designs for solar cookers available on the Internet. Have groups of students research, build, and test their own solar cookers. Consider organizing a contest to see which group can build the most efficient solar cooker.

Safety Caution: Warn students not to stare at reflected sunlight for long periods of time.

REAL-WORLD CONNECTION

Invite an expert in the field of alternative energy to speak with the class. If this is not possible, have students write a short report about an alternative energy invention of their choice.

CROSS-DISCIPLINARY FOCUS

History Humans have harnessed the energy of the wind to perform mechanical tasks for more than 1,000 years. Windmills were common in ancient China and Persia and were used primarily for grinding grain and pumping water. Windmills were introduced in Europe in the twelfth century, and by the eighteenth century, they reached levels of incredible sophistication. The invention of coal-powered steam engines in the 1760s caused the demise of many windmills. But windmills remained the most efficient way to pump water—they played a key role in the settlement of the American West. In fact, many ranchers still use windmills to pump water for cattle and irrigation. Challenge students to work in groups to make model windmills for a class competition.

BRAIN FOOD

Generating electrical power from wind energy has evolved from an experimental technology to a practical energy source: revenues from wind-generated power projects exceed $1 billion annually. Wind turbines in the United States currently produce about 3 billion kilowatt-hours of electricity a year. According to the National Wind Technology Center, the United States has the ability to use the wind to produce more electricity than our country currently uses! North Dakota alone experiences enough steady wind to supply about a third of the nation's electricity needs.

Capture the Wind

Wind is created indirectly by solar energy through the uneven heating of air. There is a tremendous amount of energy in wind, called **wind energy.** You can see the effects of this energy unleashed in a hurricane or tornado. Wind energy can also be used productively by humans. Wind energy can turn a windmill that pumps water or produces electricity.

Wind Turbines Today, fields of modern wind turbines—technological updates of the old windmills—generate significant amounts of electricity. Clusters of these turbines are often called wind farms. Wind farms are located in areas where winds are strong and steady. Most of the wind farms in the United States are in California. The amount of energy produced by California wind farms could power all of the homes in San Francisco.

Steady Breezes There are many benefits of using wind energy. Wind energy is renewable. Wind farms can be built in only 3–6 months. Wind turbines produce no carbon dioxide or other air pollutants during operation. The land used for wind farms can also be used for other purposes, such as cattle grazing, as shown in **Figure 23.** However, the wind blows strongly and steadily enough to produce electricity on a large scale only in certain places. Currently, wind energy accounts for only a small percentage of the energy used in the United States.

Figure 23 *Wind turbines take up only a small part of the ground's surface. This allows the land on wind farms to be used for more than one purpose.*

internetconnect

SCILINKS
NSTA

TOPIC: Renewable Resources
GO TO: www.scilinks.org
*sci***LINKS NUMBER:** HSTE110

SECTION REVIEW

1. Briefly describe two ways of using solar energy.

2. In addition to multiple turbines, what is needed to produce electricity from wind energy on a large scale?

3. **Analyzing Methods** Nuclear power plants are rarely found in the middle of deserts or other extremely dry areas. If you were going to build a nuclear plant, why would you not build it in the middle of a desert?

▼ *Answers to Section Review*

1. Answers will vary. A solar cell converts solar energy directly into electricity. Solar collectors convert solar energy into thermal energy.

2. A strong, steady wind is needed to produce electricity from wind energy on a large scale.

3. Large amounts of water are needed to cool nuclear plants so that they don't overheat. Deserts and other dry areas may not have abundant water supplies.

Hydroelectric Energy

The energy of falling water has been used by humans for thousands of years. Water wheels, such as the one shown in **Figure 24,** have been around since ancient times. In the early years of the Industrial Revolution, water wheels provided energy for many factories. More recently, the energy of falling water has been used to generate electricity. Electricity produced by falling water is called **hydroelectric energy.**

Harnessing the Water Cycle Hydroelectric energy is inexpensive and produces little pollution, and it is renewable because water constantly cycles from the ocean to the air, to the land, and back to the ocean. But like wind energy, hydroelectric energy is not available everywhere. Hydroelectric energy can be produced only where large volumes of falling water can be harnessed. Huge dams, like the one in **Figure 25,** must be built on major rivers to capture enough water to generate significant amounts of electricity.

Figure 24 *Falling water turns water wheels, which turn giant millstones used to grind grain into flour.*

Figure 25 *Falling water turns huge turbines inside hydroelectric dams, generating electricity for millions of people.*

At What Price? Increased use of hydroelectric energy could reduce the demand for fossil fuels, but there are trade-offs. Construction of the large dams necessary for hydroelectric power plants often destroys other resources, such as forests and wildlife habitats. For example, hydroelectric dams on the Lower Snake and Columbia Rivers in Washington disrupt the migratory paths of local populations of salmon and steelhead. Large numbers of these fish die each year because their life cycle is disrupted. Dams can also decrease water quality and create erosion problems.

> **✓ Self-Check**
>
> How are ancient water wheels like modern hydroelectric dams? *(See page 168 to check your answer.)*

CONNECT TO PHYSICAL SCIENCE

Both hydroelectric dams and wind turbines supply electricity, but how? They use a device called a *generator* to convert kinetic energy (the energy of motion) into electrical energy (the energy of moving electrons). A generator contains a large magnet that spins rapidly within thousands of tiny wire coils. As the magnet spins, electric current is generated in the wire coils. Students can demonstrate this by wrapping 1 m of thin-gauge wire around a cardboard tube. Have them attach both ends of the wire to a galvanometer and pass a bar magnet quickly through the tube.

Multicultural CONNECTION

With 70 percent of people in developing countries burning wood and charcoal for heating and cooking fuel, the use of alternative biomass fuels is crucial to slowing rates of deforestation. In the Bolivian highlands of South America, the llama is a traditional beast of burden, but its dung is even more important as a source of biomass fuel. In India, cow dung is used in many rural areas for heating and cooking fuel. In the deserts of Arabia, the nomadic Bedouin use camel dung as fuel, and many people in Tibet and Nepal use yak dung as fuel. Dried yak dung is the main fuel available on the treeless Tibetan plateaus. In China and other countries, methane produced from composting pig manure is used to generate electricity.

MEETING INDIVIDUAL NEEDS

Writing **Advanced Learners**
Have students research and prepare reports on biomass fuels. Students might choose to research the use of lumber industry wastes, agricultural wastes, organic municipal wastes, food processing wastes, aquatic plants and algae, and municipal sewage.

Answers to MATHBREAK

2.5 acres

 Critical Thinking Worksheet "Nature's Gold"

 Interactive Explorations CD-ROM "The Generation Gap"

Powerful Plants

Plants are similar to solar collectors, absorbing energy from the sun and storing it for later use. Leaves, wood, and other parts of plants contain the stored energy. Even the dung of plant-grazing animals is high in stored energy. These sources of energy are called biomass. **Biomass** is organic matter that contains stored energy.

Burning Biomass Biomass energy can be released in several ways. The most common is the burning of biomass. Approximately 70 percent of people living in developing countries heat their homes and cook their food by burning wood or charcoal. In the United States this number is about 5 percent. Scientists estimate that the burning of wood and animal dung accounts for approximately 14 percent of the world's total energy use.

Figure 26 *In many parts of the world where firewood is scarce, people burn animal dung for energy. This woman is preparing cow dung that will be dried and used as fuel.*

MATH BREAK

Miles per Acre

Imagine that you own a car that runs on alcohol made from corn that you grow. You drive your car about 15,000 miles in a year, and you get 240 gallons of alcohol from each acre of corn that you process. If your car gets 25 mi/gal, how many acres of corn would you have to grow to fuel your car for a year?

Gasohol Plant material can also be changed into liquid fuel. Plants containing sugar or starch, for example, can be made into alcohol. The alcohol is burned as a fuel or mixed with gasoline to make a fuel mixture called **gasohol.** An acre of corn can produce more than 1,000 L of alcohol. But in the United States we use a lot of fuel for our cars. It would take about 40 percent of the entire United States corn harvest to produce enough alcohol to make just 10 percent of the fuel we use in our cars! Biomass is obviously a renewable source of energy, but producing biomass requires land that could be used for growing food.

114

Water at the ocean's surface can be 25°C warmer than water 1,000 m below. Ocean Thermal Energy Conversion (OTEC) uses this temperature difference to produce electricity. Some systems use an ammonia-filled pipe that runs in a closed loop deep into the ocean. In deep, cold water, the fluid is liquid, but as it rises through the warm surface water, it becomes a gas. This expansion drives a turbine that generates electricity. The gas condenses, and the cycle repeats. Open-cycle OTEC plants, which can generate as much as 210 kW, use sea water as the condensing fluid and produce desalinized water as a byproduct.

Deep Heat

Imagine being able to tap into the energy of the Earth. In a few places this is possible. This type of energy is called geothermal energy. **Geothermal energy** is energy produced by heat within the Earth's crust.

Geothermal Energy In some locations, rainwater penetrates porous rock near a source of magma. The heat from the magma heats the water, often turning it to steam. The steam and hot water escape through natural vents called geysers, or through wells drilled into the rock. The steam and water contain geothermal energy. Some geothermal power plants use primarily steam to generate electricity. This process is illustrated in **Figure 27.** In recent years, geothermal power plants that use primarily hot water instead of steam have become more common.

Geothermal energy can also be used as a direct source of heat. In this process, hot water and steam are used to heat a fluid that is pumped through a building in order to heat the building. Buildings in Iceland are heated in this way from the country's many geothermal sites.

Figure 27 How a Geothermal Power Plant Works

④ The steam escapes the power plant through vents.

⑤ Left over liquid water is injected back into the hot rock.

③ The generators produce electricity.

② The steam drives turbines, which in turn drive electric generators.

① Steam rises through a well.

Heated water

Hot rock

SECTION REVIEW

1. Where is the production of hydroelectric energy practical?

2. Name two ways to release biomass energy.

3. Describe two ways to use geothermal energy.

4. **Summarizing Data** List four energy alternatives to fossil fuels, and give one advantage and one disadvantage of each alternative.

▼ **Answers to Section Review**

1. Hydroelectricity is practical in areas where there are large volumes of moving water.

2. Answers may vary. Methods given in text are: burning biomass and converting plant material to alcohol that can be burned.

3. Steam and hot water heated by geothermal energy can be used to generate electricity. Steam and water heated by geothermal energy can also be used as a direct heat source.

4. Answers will vary. Any four of the six alternative energy resources featured in this section are acceptable, as are reasonable alternatives not mentioned in this section.

3 Extend

GOING FURTHER

Encourage students to find out more about an alternative energy source of their choice. Then challenge them to create an accurate model showing how the energy source is used and present it to the class.

4 Close

Quiz

1. Compare and contrast fission and fusion. (Both are nuclear reactions that generate vast amounts of energy. Fission, which is presently used to generate energy, is the splitting of an atomic nucleus. Fusion is the joining of atomic nuclei.)

2. How are plants used to produce energy? (Burning wood, crops, and alcohol made from plants are some ways plants are used to produce energy.)

3. Where in the United States is geothermal energy a good alternative? (California, Hawaii, and the Yellowstone National Park area)

ALTERNATIVE ASSESSMENT

Have students compile a table that lists each fuel mentioned in this chapter and compares the advantages and disadvantages of each. As an alternative, have students apply what they have learned in this chapter to compile a list of suggestions for making their home, school, or classroom less wasteful of natural resources.

115

Make a Water Wheel
Teacher's Notes

Time Required

One or two 45-minute class periods

Lab Ratings

EASY ———————→ HARD

TEACHER PREP 🔬

STUDENT SET-UP 🔬🔬🔬

CONCEPT LEVEL 🔬

CLEAN UP 🔬🔬

The materials listed are best for a group of 3–4 students.

Safety Caution

Remind students to review all safety cautions and icons before beginning this lab activity.

Preparation Notes

One week before the activity, have students bring in empty, 1 gal plastic milk and distilled-water jugs. (Be sure that the milk jugs are thoroughly rinsed.) You can also get these from a plastic recycling drop-off center. Do not use plastic jugs that once contained harmful chemicals. Also have students bring in empty 2 L soda bottles. Obtain corks from a craft store, or collect them yourself. Some restaurants will save corks for you from opened wine bottles if you ask ahead of time. You may wish to have extra corks on hand; some corks become dry, brittle, and crumbly. Skewers can be obtained at a grocery store; they usually come in packages of 200 and are inexpensive. Pick off the rough fibers from the skewers to reduce the friction.

Make a Water Wheel

Lift Enterprises is planning to build a water wheel that will lift objects like a crane does. City planners feel that this would make very good use of the energy supplied by the river that flows through town. Development of the water wheel is in the early stages. The president of the company has asked you to modify the basic water-wheel design so that the final product will lift objects more quickly.

MATERIALS

🥽 ✂️ 🔦

- index card
- metric ruler
- scissors
- safety razor (for teacher)
- large plastic milk jug
- permanent marker
- 5 thumbtacks
- cork
- glue
- 2 wooden skewers
- hole punch
- modeling clay
- transparent tape
- 20 cm of thread
- coin
- 2 L bottle filled with water
- watch or clock that indicates seconds

116

Ask a Question

1 What factors influence the rate at which a water wheel lifts a weight?

Form a Hypothesis

2 In your ScienceLog, change the question above into a statement, giving your "best guess" as to what factors will have the greatest effect on your water wheel.

Build a Model

3 Measure and mark a 5 cm × 5 cm square on an index card. Cut the square out of the card.

4 Fold the square in half to form a triangle.

5 Measure and mark a line 8 cm from the bottom of the plastic jug. Use scissors to cut along this line. (Your teacher may need to use a safety razor to start this cut for you.) Keep both sections.

6 Use the permanent marker to trace four triangles onto the flat parts of the top section of the plastic jug. Use the paper triangle you made in step 4 as a template. Cut the triangles out of the plastic to form four fins.

Lab Notes

If the coin is lowered instead of raised in step 13, instruct students to unwrap the thread, wrap it in the other direction around the clay, and repeat step 13. You may wish to have a class competition to see which wheel can lift the weight the fastest.

CLASSROOM TESTED & APPROVED

Tracy Jahn
Berkshire Jr.–Sr. High School
Canaan, New York

7 Use a thumbtack to attach one corner of each plastic fin to the round edge of the cork, as shown on page 116. Make sure the fins are equally spaced around the cork.

8 Place a drop of glue on one end of each skewer. Insert the first skewer into one of the holes in the end of the cork. Insert the second skewer into the hole in the other end.

9 Use a hole punch to carefully punch two holes in the bottom section of the plastic jug. Punch each hole 1 cm from the top edge of the jug, directly across from one another.

10 Carefully push the skewers through the holes, and suspend the cork in the center of the jug. Attach a small ball of clay to the end of each skewer. The balls should be the same size.

11 Tape one end of the thread to one skewer on the outside of the jug, next to the clay ball. Wrap the thread around the clay ball three times. (As the water wheel turns, the thread should continue to wrap around the clay. The other ball of clay balances the weight and helps to keep the water wheel turning smoothly.)

12 Tape the free end of the thread to a coin. Wrap the thread around the coin once, and tape it again.

Test the Hypothesis

13 Slowly and carefully pour water from the 2 L bottle onto the fins so that the water wheel spins. What happens to the coin? Record your observations in your ScienceLog.

14 Lower the coin back to the starting position. Add more clay to the skewer to increase the diameter of the wheel. Repeat step 13. Did the coin rise faster or slower this time?

15 Lower the coin back to the starting position. Modify the shape of the clay, and repeat step 13. Does the shape of the clay affect how quickly the coin rises? Explain your answer.

16 What happens if you remove two of the fins from opposite sides? What happens if you add more fins? Modify your water wheel to find out.

17 Experiment with another fin shape. How does a different fin shape affect how quickly the coin rises?

Analyze the Results

18 What factors influence how quickly you can lift the coin?

Draw Conclusions

19 What recommendations would you make to Lift Enterprises to improve its water wheel?

Answers

13. The coin rises.

14. The coin rises faster with more clay.

15. If the clay is shaped so that the thread has to wrap around a bulge, the coin will rise faster. If the clay is shaped so that the thread has to wrap around a narrow part, the coin will rise slower.

16. Fewer fins cause the wheel to turn slower, and the coin rises slower. More fins cause the wheel to turn faster, and the coin rises faster.

17. Generally, fins that catch more water will cause the wheel to turn faster, and the coin will rise faster.

18. The shape and amount of clay and the number and shape of the fins influence how quickly the wheel lifts the coin.

19. Answers will vary. Recommendations could include the following: more fins, fin shapes that catch more water, and wrapping the rope or cable around a large diameter.

Datasheets for LabBook

117

Chapter Highlights

Chapter Highlights

VOCABULARY DEFINITIONS

SECTION 1

natural resource any natural substance, organism, or energy form that living things use

renewable resource a natural resource that can be used and replaced over a relatively short time

nonrenewable resource a natural resource that cannot be replaced or that can be replaced only over thousands or millions of years

recycling the process by which used or discarded materials are treated for reuse

SECTION 2

energy resource a natural resource that humans use to produce energy

fossil fuel a nonrenewable energy resource that forms in the Earth's crust over millions of years from the buried remains of once-living organisms

petroleum an oily mixture of flammable organic compounds from which liquid fossil fuels and other products are separated; crude oil

natural gas a gaseous fossil fuel

coal a solid fossil fuel formed underground from buried, decomposed plant material

strip mining a process in which rock and soil are stripped from the Earth's surface to expose the underlying materials to be mined

acid precipitation precipitation that contains acids due to air pollution

smog a photochemical fog produced by the reaction of sunlight and air pollutants

SECTION 1

Vocabulary
  natural resource *(p. 98)*
  renewable resource *(p. 99)*
  nonrenewable resource *(p. 99)*
  recycling *(p. 100)*

Section Notes

• Natural resources include everything that is not made by humans and that can be used by organisms.

• Renewable resources, like trees and water, can be replaced in a relatively short period of time.

• Nonrenewable resources cannot be replaced, or they take a very long time to replace.

• Recycling reduces the amount of natural resources that must be obtained from the Earth.

SECTION 2

Vocabulary
  energy resource *(p. 101)*
  fossil fuel *(p. 101)*
  petroleum *(p. 101)*
  natural gas *(p. 102)*
  coal *(p. 102)*
  strip mining *(p. 105)*
  acid precipitation *(p. 106)*
  smog *(p. 107)*

Section Notes

• Fossil fuels, including petroleum, natural gas, and coal, form from the buried remains of once-living organisms.

• Petroleum and natural gas form mainly from the remains of microscopic sea life.

• Coal forms from decayed swamp plants and varies in quality based on its percentage of carbon.

• Petroleum and natural gas are obtained through drilling, while coal is obtained through mining.

• Obtaining and using fossil fuels can cause many environmental problems, including acid precipitation, water pollution, and smog.

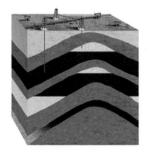

☑ Skills Check

Math Concepts

THE CARBON CONTENT OF COAL Turn back to page 104 to study the process of coal formation. Notice that at each stage, 10% more of the organic material becomes carbon. To calculate the percentage of carbon present at the next stage, just add 10%, or 0.10. For example:

peat → lignite
60% → 70%
0.60 + 0.10 = 0.70, or 70%

Visual Understanding

NO DIRECT CONTACT Take another look at Figure 21 on page 111. It is important to realize that the heated liquid inside the solar collector's tubes never comes in direct contact with the water in the tank. Cold water enters the tank, receives energy from the hot, coiled tube, and leaves the tank when someone turns on the hot-water tap.

Lab and Activity Highlights

Make a Water Wheel PG 116

Power of the Sun PG 136

 Datasheets for LabBook
(blackline masters for these labs)

SECTION 3

Vocabulary

nuclear energy *(p. 108)*
solar energy *(p. 109)*
wind energy *(p. 112)*
hydroelectric energy *(p. 113)*
biomass *(p. 114)*
gasohol *(p. 114)*
geothermal energy *(p. 115)*

Section Notes

- Nuclear energy is most often produced by fission.
- Radioactive wastes and the threat of overheating in nuclear power plants are among the major problems associated with using nuclear energy.

- Solar energy can be converted to electricity by using solar cells.
- Solar energy can be used for direct heating by using solar collectors.
- Solar energy can be converted to electricity on both a small and large scale.
- Although harnessing wind energy is practical only in certain areas, the process produces no air pollutants, and land on wind farms can be used for more than one purpose.
- Hydroelectric energy is inexpensive, renewable, and produces little pollution. However, hydroelectric dams can damage wildlife habitats, create erosion problems, and decrease water quality.

- Plant material and animal dung that contains plant material can be burned to release energy.
- Some plant material can be converted to alcohol. This alcohol can be mixed with gasoline to make a fuel mixture called gasohol.
- Geothermal energy can be harnessed from hot, liquid water and steam that escape through natural vents or through wells drilled into the Earth's crust. This energy can be used for direct heating or can be converted to electricity.

Labs

Power of the Sun *(p. 136)*

VOCABULARY DEFINITIONS, *continued*

SECTION 3

nuclear energy the form of energy associated with changes in the nucleus of an atom; an alternative energy resource

solar energy energy from the sun

wind energy energy in wind

hydroelectric energy electricity produced by falling water

biomass organic matter, such as plants, wood, and waste, that contains stored energy

gasohol a mixture of gasoline and alcohol that is burned as a fuel

geothermal energy energy from within the Earth

Vocabulary Review Worksheet

Blackline masters of these Chapter Highlights can be found in the **Study Guide.**

---

internet**connect**

go hrw .com

GO TO: go.hrw.com

Visit the **HRW** Web site for a variety of learning tools related to this chapter. Just type in the keyword:

KEYWORD: HSTENR

SCI LINKS SM
N S T A

GO TO: www.scilinks.org

Visit the **National Science Teachers Association** on-line Web site for Internet resources related to this chapter. Just type in the *sci*LINKS number for more information about the topic:

TOPIC: Natural Resources	***sci*LINKS NUMBER:** HSTE105
TOPIC: Renewable Resources	***sci*LINKS NUMBER:** HSTE110
TOPIC: Nonrenewable Resources	***sci*LINKS NUMBER:** HSTE115
TOPIC: Fossil Fuels	***sci*LINKS NUMBER:** HSTE120
TOPIC: Nuclear Energy	***sci*LINKS NUMBER:** HSTE122

119

Lab and Activity Highlights

LabBank

Calculator-Based Labs, Solar Homes

Long-Term Projects & Research Ideas, Build a City–Save a World!

Interactive Explorations CD-ROM

CD 1, Exploration 6, "The Generation Gap"

Chapter Review
Answers

USING VOCABULARY

1. An energy resource is a resource that humans can use to produce energy, while a natural resource is any natural substance, organism, or energy form that living things use.
2. Acid precipitation is formed from the mixing of air pollutants with precipitation, while smog is produced by the action of sunlight on air pollutants.
3. Biomass is organic matter that contains stored energy, while gasohol is a fuel made from plant biomass.
4. Hydroelectric energy is harnessed from falling water, while geothermal energy is harnessed from steam and hot liquid water escaping from the Earth's crust.

UNDERSTANDING CONCEPTS

Multiple Choice
5. b
6. b
7. b
8. c
9. d
10. a
11. a
12. d
13. c
14. c

Short Answer
15. It is important to conserve natural resources so that we won't use them up faster than they can be replaced.
16. Air pollutants mix with moisture in the air, producing acids. These acids fall with rain or snow as acid precipitation.
17. Solar cells are relatively expensive to make.

Chapter Review

USING VOCABULARY

For each pair of terms, explain the difference in their meanings.

1. natural resource/energy resource
2. acid precipitation/smog
3. biomass/gasohol
4. hydroelectric energy/ geothermal energy

UNDERSTANDING CONCEPTS

Multiple Choice

5. Of the following, the one that is a renewable resource is
   a. coal.
   b. trees.
   c. oil.
   d. natural gas.

6. All of the following are separated from petroleum except
   a. jet fuel.
   b. lignite.
   c. kerosene.
   d. fuel oil.

7. Which of the following is a component of natural gas?
   a. gasohol
   b. methane
   c. kerosene
   d. gasoline

8. Peat, lignite, and anthracite are all stages in the formation of
   a. petroleum.
   b. natural gas.
   c. coal.
   d. gasohol.

9. Which of the following factors contribute to smog problems?
   a. high numbers of automobiles
   b. lots of sunlight
   c. mountains surrounding urban areas
   d. all of the above

10. Which of the following resources produces the least pollution?
    a. solar energy
    b. natural gas
    c. nuclear energy
    d. petroleum

11. Nuclear power plants use a process called _____?_____ to produce energy.
    a. fission
    b. fusion
    c. fractionation
    d. None of the above

12. A solar-powered calculator uses
    a. solar collectors.
    b. solar panels.
    c. solar mirrors.
    d. solar cells.

13. Which of the following is a problem with using wind energy?
    a. air pollution
    b. amount of land required for wind turbines
    c. limited locations for wind farms
    d. none of the above

14. Dung is a type of
    a. geothermal energy.
    b. gasohol.
    c. biomass.
    d. None of the above

Short Answer

15. Because renewable resources can be replaced, why do we need to conserve them?

16. How does acid precipitation form?

17. If sunlight is free, why is electricity from solar cells expensive?

Concept Mapping

18. Use the following terms to create a concept map: fossil fuels, wind energy, energy resources, biomass, renewable resources, solar energy, nonrenewable resources, natural gas, gasohol, coal, oil.

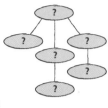

CRITICAL THINKING AND PROBLEM SOLVING

Write one or two sentences to answer the following questions:

19. How would your life be different if all fossil fuels suddenly disappeared?

20. Are fossil fuels really nonrenewable? Explain.

21. What solutions are there for the problems associated with nuclear waste?

22. How could the problems associated with the dams in Washington and local fish populations be solved?

23. What limits might there be on the productivity of a geothermal power plant?

MATH IN SCIENCE

24. Imagine that you are designing a solar car. If you mount solar cells on the underside of the car as well as on the top in direct sunlight, and it takes five times as many cells underneath to generate the same amount of electricity generated by the cells on top, what percentage of the sunlight is reflected back off the pavement?

INTERPRETING GRAPHICS

The chart below shows how various energy resources meet the world's energy needs. Use the chart to answer the following questions:

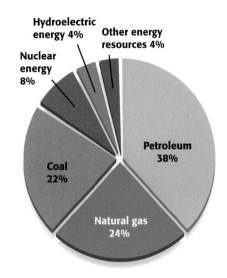

Hydroelectric energy 4%
Other energy resources 4%
Nuclear energy 8%
Petroleum 38%
Coal 22%
Natural gas 24%

25. What percentage of the world's total energy needs is met by coal? by natural gas? by hydroelectric energy?

26. What percentage of the world's total energy needs is met by fossil fuels?

27. How much more of the world's total energy needs is met by petroleum than by natural gas?

Reading Check-up

Take a minute to review your answers to the Pre-Reading Questions found at the bottom of page 96. Have your answers changed? If necessary, revise your answers based on what you have learned since you began this chapter.

121

Concept Mapping Transparency 5

Blackline masters of this Chapter Review can be found in the **Study Guide.**

Concept Mapping

18. An answer to this exercise can be found at the front of this book.

CRITICAL THINKING AND PROBLEM SOLVING

19. Answers will vary. Answers should discuss using alternative energy resources.

20. Answers will vary. We label certain resources as nonrenewable because it takes a long time for them to be replenished. Fossil fuels (and many other nonrenewable resources) are renewable in the sense that they will probably be renewed in the future. But they cannot be renewed in time for us to count on them as energy resources.

21. Answers will vary. Answers may discuss storing nuclear waste in safe areas or minimizing or eliminating the use of nuclear energy to avoid producing more radioactive waste.

22. Answers will vary. Answers may discuss modifying the dams to allow fish populations to migrate, decreasing erosion, and improving water quality. Students may even suggest moving the dams to an area where migratory fish populations would not be affected. Encourage students to research these problems at a library or on the Internet.

23. Answers will vary. Students may discuss the amount of geothermal energy available or the number of vents the plant uses.

MATH IN SCIENCE

24. 20%

INTERPRETING GRAPHICS

25. 22%; 24%; 4%
26. 84%
27. 14%

Chapter 5 • Chapter Review **121**

Background

Petroleum is not important just as a fuel source. It also plays several vital roles in the production and recycling of plastics. First, petroleum is the main component of plastics. Second, petroleum is often used as fuel to provide the energy used in the manufacture of plastic products. If these products are recycled, the melting and reshaping processes require heat and electricity, which may come from the burning of petroleum fuels.

Teaching Strategy

Does your school recycle? If not, work with students to find out how to start a recycling program in your school. This would be a good opportunity to work with other schools in your district by setting up a districtwide program if you don't already have one. If you do have a recycling program, do some research to find out how much is recycled at your school and other schools in your district. Encourage students to recycle at home if they don't already.

EYE ON THE ENVIRONMENT

Sitting on Your Trash

Did you know that the average person creates about 2 kg of waste every day? About 7 percent of this waste is composed of plastic products that can be recycled. Instead of adding to the landfill problem, why not recycle your plastic trash so you can sit on it? Well you can, you know! Today plastic is recycled into products like picnic tables, park benches, and even highchairs! But how on Earth does the plastic you throw away become a park bench?

Sort It Out

Once collected and taken to a recycling center, plastic must be sorted. This process involves the coded symbols that are printed on every recyclable plastic product we use. Each product falls into one of two types of plastic—*polyethylene* or *polymer*. The plastic mainly used to make furniture includes the polyethylene plastics called *high density polyethylene,* or HDPE, and *low density polyethylene,* or LDPE. These are items such as milk jugs, detergent bottles, plastic bags, and grocery bags.

Grind It and Wash It

The recycling processes for HDPE and LDPE are fairly simple. Once it reaches the processing facility, HDPE plastic is ground into small flakes about 1 cm in diameter. In the case of LDPE plastic, which are thin films, a special grinder is used to break it down. From that point on, the recycling process is pretty much the same for LDPE and HDPE. The pieces are then washed with hot water and detergent. In this step, dirt and things like labels are removed. After the wash, the flakes are dried with blasts of hot air.

Recycle It!

Some recycling plants sell the recycled flakes. But others may reheat the flakes, change the color by adding a pigment, and then put the material in a *pelletizer*. The little pellets that result are then purchased by a company that molds the pellets into pieces of plastic lumber. This plastic lumber is used to create flowerpots, trash cans, pipes, picnic tables, park benches, toys, mats, and many other products!

From waste…

to plastic lumber…

to a park bench!

Can You Recycle It?

▶ The coded symbol on a plastic container tells you what type of plastic the item is made from, but it doesn't mean that you can recycle it in your area. Find out which plastics can be recycled in your state.

122

Answer to Can You Recycle It?
Answers will vary.

Eureka!

Oil Rush!

You may have heard of the great California gold rush. In 1849, thousands of people moved to the West hoping to strike gold. But you may not have heard about another rush that followed 10 years later. What lured people to northwestern Pennsylvania in 1859? The thrill of striking oil!

Demand for Petroleum

People began using oil as early as 3000 B.C., and oil has been a valuable substance ever since. In Mesopotamia, people used oil to waterproof their ships. The Egyptians and Chinese used oil as a medicine. It was not until the late 1700s and early 1800s that people began to use oil as a fuel. Oil was used to light homes and factories.

Petroleum Collection

But what about the oil in northwestern Pennsylvania? Did people use the oil in Pennsylvania before the rush of 1859? Native Americans were the first to dig pits to collect oil near Titusville, Pennsylvania. Early settlers used the oil as a medicine and as a fuel to light their homes. But their methods for collecting the oil were very inefficient.

The First Oil Well

In 1859, "Colonel" Edwin L. Drake came up with a better method of collecting oil from the ground. Drilling for oil! Drake hired salt-well drillers to burrow to the bedrock where oil deposits lay. But each effort was unsuccessful because water seeped into the wells, causing them to cave in. Then Drake came up with a unique idea that would make him a very wealthy man. Drake suggested that the drillers drive an iron pipe down to the bedrock 21.2 m below the surface. Then they could drill through the inner diameter of the pipe. The morning after the iron pipe was drilled, Drake woke to find that the pipe had filled with oil!

Oil City

Within 3 months, nearly 10,000 people rushed to Oil City, Pennsylvania, in search of the wealth that oil promised. Within 2 years, the small village became a bustling oil town of 50,000 people! In 1861, the first gusher well was drilled nearby, and some 3,000 barrels of oil spouted out daily. Four years later, the first oil pipeline carried crude oil a distance of 8 km.

▲ *Edwin Drake (right) and his friend Peter Wilson (left) in front of Drake Oil Well, near Titusville, Pennsylvania*

Find Out for Yourself!

▶ Drake's oil well was the first well used to collect oil from the ground. Research the oil wells today. How are they similar to Drake's well?

123

Answer to Find Out for Yourself!

Today's oil wells are somewhat similar to Drake's oil well in that they also use a metal pipe. First a hole, or *well bore*, is drilled into the ground. The equivalent of Drake's metal pipe is a metal casing that is inserted into the well bore. Cement is pumped into the hole and fills the narrow space between the well bore and the casing. Once the cement dries, the casing is bonded to the well bore and prevents the contamination of oil, gas, and water resources that otherwise might flow through the oil well.

Background

Until 1880, people in the United States relied on vegetable and animal oils to light their homes. The production of oil products from these sources was time intensive and costly. Whales, in particular, became rare because they were hunted for their oils during the nineteenth century.

Drake's well ushered in the modern era of the petroleum industry. His drilling methods allowed oil to be collected quickly and inexpensively. By 1900, nearly 64 million barrels of oil had been collected in the United States. Many of Drake's original techniques have been adapted and are still being used today.

In 1896, the first offshore oil drilling operations were started off the coast of California. The operations opened new and potentially lucrative opportunities to oil speculators. By 1938, the first oil platform had been built off the Louisiana coast.

Some experts wonder whether we will face an oil shortage in the near future. Fewer and fewer oil reserves are being discovered. Some people wonder if the recently discovered oil deposits in the Caspian Sea may be the site of the last great oil rush.

Encourage students to consider alternative sources of energy. For example, they may want to investigate the use of solar energy to power cars and heat homes.

Exploring, inventing, and investigating are essential to the study of science. However, these activities can also be dangerous. To make sure that your experiments and explorations are safe, you must be aware of a variety of safety guidelines.

You have probably heard of the saying, "It is better to be safe than sorry." This is particularly true in a science classroom where experiments and explorations are being performed. Being uninformed and careless can result in serious injuries. Don't take chances with your own safety or with anyone else's.

Following are important guidelines for staying safe in the science classroom. Your teacher may also have safety guidelines and tips that are specific to your classroom and laboratory. Take the time to be safe.

Safety Rules!

Start Out Right

Always get your teacher's permission before attempting any laboratory exploration. Read the procedures carefully, and pay particular attention to safety information and caution statements. If you are unsure about what a safety symbol means, look it up or ask your teacher. You cannot be too careful when it comes to safety. If an accident does occur, inform your teacher immediately, regardless of how minor you think the accident is.

Safety Symbols

All of the experiments and investigations in this book and their related worksheets include important safety symbols to alert you to particular safety concerns. Become familiar with these symbols so that when you see them, you will know what they mean and what to do. It is important that you read this entire safety section to learn about specific dangers in the laboratory.

If you are instructed to note the odor of a substance, wave the fumes toward your nose with your hand. Never put your nose close to the source.

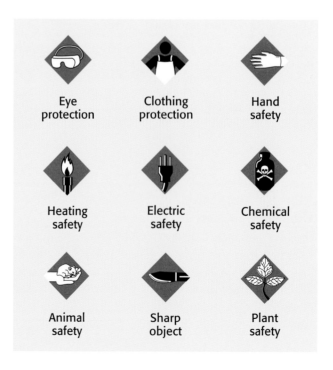

Eye protection

Clothing protection

Hand safety

Heating safety

Electric safety

Chemical safety

Animal safety

Sharp object

Plant safety

Eye Safety

Wear safety goggles when working around chemicals, acids, bases, or any type of flame or heating device. Wear safety goggles any time there is even the slightest chance that harm could come to your eyes. If any substance gets into your eyes, notify your teacher immediately, and flush your eyes with running water for at least 15 minutes. Treat any unknown chemical as if it were a dangerous chemical. Never look directly into the sun. Doing so could cause permanent blindness.

Avoid wearing contact lenses in a laboratory situation. Even if you are wearing safety goggles, chemicals can get between the contact lenses and your eyes. If your doctor requires that you wear contact lenses instead of glasses, wear eye-cup safety goggles in the lab.

Safety Equipment

Know the locations of the nearest fire alarms and any other safety equipment, such as fire blankets and eyewash fountains, as identified by your teacher, and know the procedures for using them.

Be extra careful when using any glassware. When adding a heavy object to a graduated cylinder, tilt the cylinder so the object slides slowly to the bottom.

Neatness

Keep your work area free of all unnecessary books and papers. Tie back long hair, and secure loose sleeves or other loose articles of clothing, such as ties and bows. Remove dangling jewelry. Don't wear open-toed shoes or sandals in the laboratory. Never eat, drink, or apply cosmetics in a laboratory setting. Food, drink, and cosmetics can easily become contaminated with dangerous materials.

Certain hair products (such as aerosol hair spray) are flammable and should not be worn while working near an open flame. Avoid wearing hair spray or hair gel on lab days.

Sharp/Pointed Objects

Use knives and other sharp instruments with extreme care. Never cut objects while holding them in your hands. Place objects on a suitable work surface for cutting.

Heat

Wear safety goggles when using a heating device or a flame. Whenever possible, use an electric hot plate as a heat source instead of an open flame. When heating materials in a test tube, always angle the test tube away from yourself and others. In order to avoid burns, wear heat-resistant gloves whenever instructed to do so.

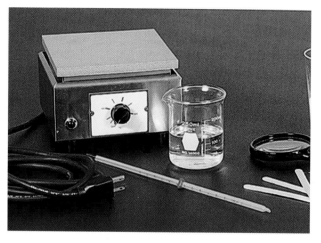

Chemicals

Wear safety goggles when handling any potentially dangerous chemicals, acids, or bases. If a chemical is unknown, handle it as you would a dangerous chemical. Wear an apron and safety gloves when working with acids or bases or whenever you are told to do so. If a spill gets on your skin or clothing, rinse it off immediately with water for at least 5 minutes while calling to your teacher.

Never mix chemicals unless your teacher tells you to do so. Never taste, touch, or smell chemicals unless you are specifically directed to do so. Before working with a flammable liquid or gas, check for the presence of any source of flame, spark, or heat.

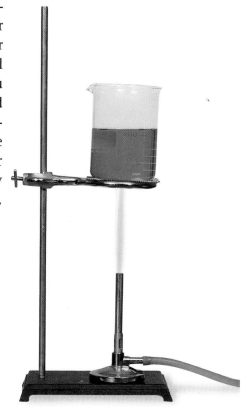

Electricity

Be careful with electrical cords. When using a microscope with a lamp, do not place the cord where it could trip someone. Do not let cords hang over a table edge in a way that could cause equipment to fall if the cord is accidentally pulled. Do not use equipment with damaged cords. Be sure your hands are dry and that the electrical equipment is in the "off" position before plugging it in. Turn off and unplug electrical equipment when you are finished.

Animal Safety

Always obtain your teacher's permission before bringing any animal into the school building. Handle animals only as your teacher directs. Always treat animals carefully and with respect. Wash your hands thoroughly after handling any animal.

Plant Safety

Do not eat any part of a plant or plant seed used in the laboratory. Wash hands thoroughly after handling any part of a plant. When in nature, do not pick any wild plants unless your teacher instructs you to do so.

Glassware

Examine all glassware before use. Be sure that glassware is clean and free of chips and cracks. Report damaged glassware to your teacher. Glass containers used for heating should be made of heat-resistant glass.

Capturing the Wild Bean

Teacher's Notes

Time Required

One 45-minute class period

Lab Ratings

EASY ———————→ HARD

TEACHER PREP 🧪🧪
STUDENT SET-UP 🧪🧪
CONCEPT LEVEL 🧪🧪
CLEAN UP 🧪

MATERIALS

The materials listed on the student page are enough for a group of 4–5 students. Large, dried beans of any kind will work well in this exercise.

Safety Caution

Remind students to review all safety cautions and icons before beginning this lab activity.

Lab Notes

Explain to students that this is a very common method used in field biology by scientists who need to count a population that is on the move, such as a flock of migratory birds. Explain that scientists always mark their captures in a way that neither harms the organism nor alters its behavior.

Capturing the Wild Bean

When wildlife biologists study a group of organisms in an area, they need to know how many organisms live there. Occasionally, biologists worry that a certain organism is outgrowing the environment's carrying capacity. Other times, scientists need to know if an organism is becoming rare so steps can be taken to protect it. However, animals can be difficult to count because they can move around and hide. Because of this, biologists have developed methods to estimate the number of animals in a specific area. One of these counting methods is called the mark-recapture method.

In this activity, you will enter the territory of the wild pinto bean to get an estimation of the number of beans that live in their paper-bag habitat.

Materials

- small paper lunch bag
- pinto beans
- permanent marker
- calculator (optional)

Procedure

1. Prepare a data table in your ScienceLog like the one below.

Mark-Recapture Data Table				
Number of animals in first capture	Total number of animals in recapture	Number of marked animals in recapture	Calculated estimate of population	Actual total population
	DO NOT WRITE IN BOOK			

2. Your teacher will provide you with a paper bag containing an unknown number of beans. Carefully reach into the bag and remove a handful of beans.

3. Count the number of beans you have "captured," and record this number in your data table under "Number of animals in first capture."

Datasheets for LabBook

Science Skills Worksheet
"Hints for Oral Presentations"

4. Use the permanent marker to carefully mark each bean that you have just counted. Allow the marks to dry completely. When you are certain that all the marks are dry, place the marked beans back into the bag.

5. Gently mix the beans in the bag so the marks won't rub off. Once again, reach into the bag, "capture," and remove a handful of beans.

6. Count the number of beans in your "recapture." Record this number in your data table under "Total number of animals in recapture."

7. Count the beans in your recapture with marks from the first capture. Record this number in your data table under "Number of marked animals in recapture."

8. Calculate your estimation of the total number of beans in the bag using the following equation:

$$\frac{\text{total number of beans in recapture} \times \text{total number of beans marked}}{\text{number of marked beans in recapture}} = \begin{array}{c}\text{calculated estimate of}\\\text{population}\end{array}$$

Enter this number in your data table under "Calculated estimate of population."

9. Replace all the beans in the bag. Then empty the bag on your work table. Be careful that no beans escape! Count each bean as you place them one at a time back into the bag. Record the number in your data table under "Actual total population."

Analysis

10. How close was your estimate to the actual number of beans?

11. If your estimate was not close to the actual number of beans, how might you change your mark-recapture procedure? If you did not recapture any marked beans, what might be the cause?

Going Further

How could you use the mark-recapture method to estimate the population of turtles in a small pond? Explain your procedure.

Answers

10. Have students evaluate their estimate. Ask them if they think their estimate was close enough to the actual number of beans in the bag.

11. Samples that are too small can lead to incorrect estimates. If no marked beans were recaptured, then the sample was too small, and increasing the sample size would improve the procedure. A sample should be large enough to allow a scientist to accurately estimate the organisms in the population without having to count every single organism.

Math Skills Worksheet
"Random Samples: Estimating Population"

Going Further

Students should state the size of the pond and form a reasonable hypothesis, including how many animals should be in the sample size. Students should describe a way to capture turtles without harming them. They should recommend a marking method that will not harm the turtles or alter the turtles' behavior.

The turtles should be returned in their environment immediately and given a day or two to recover from the trauma.

Important: students should recognize the need for the recapture size to be the same size as the original capture size.

Nitrogen Needs
Teacher's Notes

Time Required

One 45-minute class period

Lab Ratings

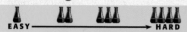
EASY —————→ HARD

TEACHER PREP — 2
STUDENT SET-UP — 3
CONCEPT LEVEL — 3
CLEAN UP — 2

MATERIALS

Dead insects may or may not be easy to come by. You may want to go to a pet store and purchase crickets. They are usually small and very inexpensive. If the insects you use are small, you will need more than five of them.

Safety Caution

Remind students to review all safety cautions and icons before beginning this lab activity. Spills should be cleaned up immediately. Hands should be washed thoroughly after all lab activities.

Preparation Notes

Caution students to collect insects carefully, wearing gardening gloves, shoes, and socks.

If you purchase insects, you can humanely kill them by placing them in a jar containing a cotton ball soaked in rubbing alcohol.

Nitrogen Needs

DISCOVERY LAB

The nitrogen cycle is one of several cycles that are vital to living organisms. Without nitrogen, living organisms cannot make amino acids, the building blocks of proteins. Animals obtain nitrogen by eating plants that contain nitrogen and by eating animals that eat those plants. When animals die, decomposers return the nitrogen to the soil in the form of a chemical called ammonia.

In this activity, you will be investigating the nitrogen cycle inside a closed system to discover how decomposers return nitrogen to the soil.

Procedure

1. Fit a piece of filter paper into a funnel. Place the funnel inside a 50 mL beaker, and pour 5 g of soil into the funnel. Add 25 mL of distilled water to the soil.

2. Test the filtered water with pH paper, and record your observations in your ScienceLog.

3. Place some soil in a jar to cover the bottom about 5 cm deep. Add 10 mL of distilled water to the soil.

4. Place the dead insects in the jar, and seal the jar with the lid.

5. Check the jar each day for 5 days for an ammonia odor. (If you do not know what ammonia smells like, ask your teacher.) Record your observations in your ScienceLog. **Caution:** Your teacher will demonstrate how to check for a chemical odor by wafting. Do not put your nose in the jar and inhale!

6. On the fifth day, place a second piece of filter paper into the funnel, and place the funnel inside a 50 mL beaker. Remove about 5 g of soil from the jar, and place it in the funnel. Add 25 mL of distilled water to the soil.

7. Once again, test the filtered water with pH paper, and record your observations in your ScienceLog.

Materials

- 2 pieces of filter paper
- funnel
- 50 mL beaker
- balance
- commercially prepared potting soil without fertilizer
- 25 mL graduated cylinder
- 60 mL of distilled water
- pH paper
- 1 pt (or 500 mL) jar with lid
- 5 large, dead insects from home or schoolyard
- protective gloves

Datasheets for LabBook

David Sparks
Redwater Junior High School
Redwater, Texas

CLASSROOM TESTED & APPROVED

Analysis

8. What was the pH of the water in the beaker in the first trial? A pH of 7 indicates that the water is neutral. A pH below 7 indicates that the water is acidic, and a pH above 7 indicates that the water is basic.

9. What was the pH of the water in the beaker in the second trial? Explain the difference, if any, between the results of the first trial and the results of the second trial.

10. Based on the results of your pH tests, do you think ammonia is acidic or basic?

11. On which days in your investigation were you able to detect an ammonia odor? Explain what caused the odor.

12. Describe the importance of decomposers in the nitrogen cycle.

Going Further

Test ammonia's importance to plants. Fill two 12 cm flowerpots with commercially prepared potting soil and water. Be sure to use soil that has had no fertilizer added. Plant six radish seeds in each pot. Water your seeds so that the soil is constantly damp but not soaked. Keep your pots in a sunny window. You may plant other seeds of your choice, but do not use legume (bean) seeds. Research to find out why!

Use a plant fertilizer mixed according to the directions on the container to fertilize one of the pots once a week. Water the other pot once a week with tap water.

After the seedlings appear, use a metric ruler to measure the growth of the plants in both pots. Measure the plants once a week, and record your results in your ScienceLog.

Dip in – read while still moist.
Immerse in weakly-buffered solutions u
there is no further colour change (1–1

| 0 | 1 | 2 | 3 | 4 | 5 | 6 | 7 | 8 | 9 | 10 | 11 | 12 | 13 | 14 |

Answers

8. The filtered water in the first trial should be slightly acidic; however, this will depend on the type of potting soil you are using. The soil pH may not fall inside the ranges shown in the chart. If not, expand the chart and explain to students that pH is a relative value.

9. If ammonia is present, the filtered water will be more basic than the filtered water in the first trial. Decomposition of the dead insects should be the only difference between the first and second trials.

10. Ammonia is basic.

11. Answers will vary. Students should understand that the odor is from bacterial conversion of nitrogen to ammonia.

12. Decomposers convert nitrogen to a form that plants can use. Nitrogen is essential for plants.

Going Further

The plants grown with ammonia should be bigger and fuller. The plants grown without ammonia may not survive. The only difference between the two pots should be nutrients.

131

Life in the Desert
Teacher's Notes

Time Required

One 45-minute class period

Lab Ratings

EASY ———————→ HARD

TEACHER PREP 🧪

STUDENT SET-UP 🧪🧪

CONCEPT LEVEL 🧪🧪

CLEAN UP 🧪

MATERIALS

The sponges used in this lab can be either natural sponges or the synthetic sponges available in grocery stores. Use 3 × 6 in. sponges, 1 per student, cut in half.

Answers

6. Students should describe the kind of covering or protection they provided for their "adapted" sponge. Effectiveness of the adaptation will be measured by the amount of water lost over 24 hours. Students will want their sponges to dry out as little as possible.

7. The unprotected sponge represents the organism that has no adaptation for conserving water. The unprotected sponge should dry out far more than the protected sponge.

James Chin
Frank A. Day Middle School
Newtonville, Massachusetts

Life in the Desert

Organisms that live in the desert have some very unusual methods for conserving water. Conserving water is an important function for all organisms that live on land, but it is a special challenge for animals that live in the desert. In this activity you will invent an "adaptation" for a desert animal, represented by a piece of sponge, to find out how much water the animal can conserve over a 24-hour period. You will protect your wet desert sponge so it will dry out as little as possible.

Materials

- 2 pieces of dry sponge (8 × 8 × 2 cm)
- water
- balance
- other materials as needed

Procedure

1. Plan a method for keeping your "desert animal" from drying out. Your "animal" must be in the open for at least 4 hours during the 24-hour period. Real desert animals often expose themselves to the dry desert heat in order to search for food. Write your plan in your ScienceLog. Write down your predictions about the outcome of your experiment.

2. Design data tables, if necessary, and draw them in your ScienceLog. Have your teacher approve your plan before you begin.

3. Soak two pieces of sponge in water until they begin to drip. Place each piece on a balance, and record its mass in your ScienceLog.

4. Immediately begin to protect one piece of sponge according to your plan. Place both of the pieces together in an area where they will not be disturbed. You may take your protected "animal" out for feeding as often as you want, for a total of at least 4 hours.

5. At the end of 24 hours, place each piece of sponge on the balance again, and record its mass in your ScienceLog.

Analysis

6. Describe the adaptation you used to help your "animal" survive. Was it effective? Explain.

7. What was the purpose of leaving one of the sponges unprotected? How did the water loss in each of your sponges compare?

Going Further

Conduct a class discussion about other adaptations and results. How can you relate these invented adaptations to adaptations for desert survival among real organisms?

132

Going Further

Ask students to consider the adaptation they designed for their sponge. Did it compare with a real-life adaptation in a desert organism? In what way? Discuss behavior adaptations for conserving water. Discuss endothermy and ectothermy, and ask students what adaptations conserve water and help regulate body heat.

 Datasheets for LabBook

Discovering Mini-Ecosystems

In your study of ecosystems you learned that a biome is a very large ecosystem that includes a set of smaller, related ecosystems. For example, a coniferous forest biome may include a river ecosystem, a wetland ecosystem, and a lake ecosystem. Each of those ecosystems may include several other smaller, related ecosystems. Even cities have mini-ecosystems! You may find a mini-ecosystem on a patch of sidewalk, in a puddle of rainwater, under a leaky faucet, in a shady area, or under a rock. In this activity, you will design a method for comparing two different mini-ecosystems found near your school.

Materials

- materials as needed for each investigation

Procedure

1. Examine the grounds around your school, and select two different areas you wish to investigate. Be sure to get your teacher's approval before you begin.

2. Decide what you want to learn about your mini-ecosystems. For example, you may want to know what kind of living things each area contains. You may want to list the abiotic factors of each mini-ecosystem.

3. For each mini-ecosystem, make data tables for recording your observations. You may choose to observe the mini-ecosystems for an hour. You may choose to observe the mini-ecosystems for a short period of time at several different times during the day or at the same time for several days. Get your plan approved by your teacher, and make the appropriate data tables.

Analysis

4. What factors determine the differences between your mini-ecosystems? Identify the factors that set each mini-ecosystem apart from its surrounding area.

5. How do the populations of your mini-ecosystems compare?

6. Identify some of the adaptations that the organisms living in your two mini-ecosystems have. Describe how the adaptations help the organisms survive in their environment.

7. Write a report describing and comparing your mini-ecosystems with those of your classmates.

 133

Discovering Mini-Ecosystems
Teacher's Notes

Time Required
One to two 45-minute class periods

Lab Ratings
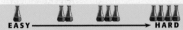

EASY ——————→ HARD

TEACHER PREP   🧪
STUDENT SET-UP   🧪🧪
CONCEPT LEVEL   🧪🧪🧪
CLEAN UP   🧪

MATERIALS
Because this is mainly an observation activity, few materials are needed. If they are available, however, binoculars or magnifying lenses may be helpful.

Lab Notes
Even if your school has no area where there is sand, dirt, grass, or trees, ask students to observe puddles, the underside of eaves, the area under drain spouts, and the ground under rocks. Students should observe the areas they have chosen at least twice a day.

Datasheets for LabBook

Barry Bishop
San Rafael Junior High
Ferron, Utah

Answers

4. Students may have many answers, but they should include answers such as different vegetation, organisms that live there, the soil type, density of vegetation, and amount of water present.

5. Students should recognize that the populations present in each area are adapted for life in that area.

6. Adaptations that students name will probably include camouflage, deep roots, and burrowing behavior.

7. Answers will vary according to student observations.

Biodiversity — What a Disturbing Though!
Teacher's Notes

Time Required
One or two 45-minute class periods

Lab Ratings

EASY ———————→ HARD

TEACHER PREP 🍼🍼

STUDENT SET-UP 🍼🍼

CONCEPT LEVEL 🍼🍼

CLEAN UP 🍼🍼

MATERIALS

Binoculars may not be available in your classroom. Ask students if they might have some at home they could get permission to bring to class for this activity.

Lab Notes
The lab should be reviewed ahead of time. This lab can be extended to a field trip that can get parents involved. Your school may be in a city where there is no suitable undisturbed area. If you are unable to take students on a field trip, help them understand the difference between severely disturbed (a paved parking lot) and an area that is less disturbed, such as an unimproved lot. Some diversity should exist in every area. You may find this lab interesting to repeat during different seasons.

DISCOVERY LAB

Biodiversity—What a Disturbing Thought!

Biodiversity is important for the survival of each organism in a community. Producers, consumers, and decomposers all play a cooperative role in an ecosystem.

In this activity you will investigate areas outside your school to determine which areas contain the greatest biodiversity. You will use the information you gather to determine whether a forest or an area planted with crops is more diverse.

Materials
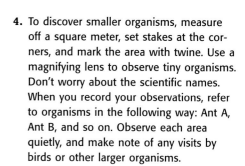

- materials and tools necessary to carry out your investigation with your teacher's approval. Possible materials include a meterstick, binoculars, magnifying lens, twine, and forceps.

Form a Hypothesis

1. Based on your understanding of biodiversity, do you expect a forest or an area planted with crops to be more diverse?

Make a Prediction

2. Select an area that is highly disturbed (such as a mowed yard) and one that is relatively undisturbed (such as an abandoned flower bed or a vacant lot). Make a prediction about which area contains the greater biodiversity. Get your teacher's approval of your selected locations.

Conduct an Experiment

3. Design a procedure to determine which area contains the greatest biodiversity, and have your plan approved by your teacher before you begin.

4. To discover smaller organisms, measure off a square meter, set stakes at the corners, and mark the area with twine. Use a magnifying lens to observe tiny organisms. Don't worry about the scientific names. When you record your observations, refer to organisms in the following way: Ant A, Ant B, and so on. Observe each area quietly, and make note of any visits by birds or other larger organisms.

CLASSROOM TESTED & APPROVED

Terry Rakes
Elmwood Junior High School
Rogers, Arkansas

5. In your ScienceLog, create any data tables that you might need for recording your data. If you observe your areas on more than one occasion, be sure to make data tables for each observation period. Organize your data into categories that are clear and understandable.

Analyze the Results

6. Did your data support your hypothesis? Explain.

7. What factors did you consider before deciding which habitats were disturbed or undisturbed? Explain why those factors were important.

8. What problems did you find in making observations and recording data for each habitat? Describe how you solved them.

9. Describe possible errors in your investigation method. Suggest ways to improve your procedure to eliminate those errors.

Draw Conclusions

10. Do you think the biodiversity outside your school has decreased since the school was built? Why or why not?

11. Both areas shown in the photographs at right are beautiful to observe. One of them, however, is very low in biodiversity. Describe each photograph, and account for the difference in biodiversity.

Going Further

Research rain-forest biodiversity in the library or on the Internet. Find out what factors exist in the rain forest that make that biome so diverse. How might the biodiversity of a rain forest compare with that of a forest community near your school?

Prairie grasses and wildflowers

Wheat field

Answers

6. Students should explain why they predicted one area would be more diverse than the other. If their predictions were confirmed, have them explain if their reasons were also correct.

7. Generally, any area inside a city or even farmland is considered ecologically disturbed. This is the conclusion you want students to come to. Students should consider human impact when deciding if an area is disturbed or undisturbed.

8. Answers will vary.

9. A possible error might be deciding that an area is undisturbed and finding that it is highly disturbed and has little diversity. A bed of petunias is lovely, but it is disturbed and not very diverse, especially if it is well weeded and insect controlled.

10. Answers will vary. Discuss construction and growth in the neighborhood since the school was built.

11. A wheat field is less diverse than a tall-grass prairie. The wheat field grows only one plant and provides little habitat for animal organisms. Some animals may visit and feed there, but their nests or dens would be in danger at harvest time.

Going Further

Have students compare a nearby forest and its diversity with a rain forest and the diversity that is found there.

 Datasheets for LabBook

 Science Skills Worksheet "Researching on the Web"

Power of the Sun

Teacher's Notes

Time Required

One or two 45-minute class periods

Lab Ratings

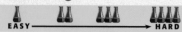
EASY ——————→ HARD

TEACHER PREP 🧪🧪
STUDENT SET-UP 🧪
CONCEPT LEVEL 🧪🧪
CLEAN UP 🧪

MATERIALS

The materials listed for this lab are enough for a group of 2–3 students.

Safety Caution

Remind students to review all safety cautions and icons before beginning this lab activity. Instruct students to not look directly at the sun. Also caution them not to squeeze the aluminum too hard around the thermometer to avoid crushing the thermometer bulb. Also tell students not to force the thermometers through the holes if the holes are too small.

Preparation Notes

Prepare the lids by punching a hole in the center of each lid with a nail and hammer. Flatten the jagged edges against the inside of the lid with the hammer. Each hole should be big enough to accommodate a thermometer. Cut 2 x 8 cm strips of aluminum from the bottom of a pie plate. Make sure there are no sharp burrs left on the edges of the strips. Before beginning this lab, you may wish to review the concept of ratios with students.

Power of the Sun

The sun radiates energy in every direction. Like the sun, the energy radiated by a light bulb spreads out in all directions. But how much energy an object receives depends on how close that object is to the source. As you move farther from the source, the amount of energy you receive decreases. For example, if you measure the amount of energy that reaches you from a light and then move three times farther away, you will discover that nine times less energy will reach you at your second position. Energy from the sun travels as light energy. When light energy is absorbed by an object it is converted into thermal energy. *Power* is the rate at which one form of energy is converted to another, and it is measured in *watts.* Because power is related to distance, nearby objects can be used to measure the power of far-away objects. In this lab you will calculate the power of the sun using an ordinary 100-watt light bulb.

Materials

- protective gloves
- aluminum strip, 2 × 8 cm
- pencil
- black permanent marker
- Celsius thermometer
- mason jar, cap, and lid with hole in center
- modeling clay
- desk lamp with a 100 W bulb and removable shade
- metric ruler
- watch or clock that indicates seconds
- scientific calculator

Procedure

1. Gently shape the piece of aluminum around a pencil so that it holds on in the middle and has two wings, one on either side of the pencil.

2. Bend the wings outward so that they can catch as much sunlight as possible.

3. Use the marker to color both wings on one side of the aluminum strip black.

4. Remove the pencil and place the aluminum snugly around the thermometer near the bulb.
   Caution: Do not press too hard—you do not want to break the thermometer! Wear protective gloves when working with the thermometer and the aluminum.

5. Carefully slide the top of the thermometer through the hole in the lid. Place the lid on the jar so that the thermometer bulb is inside the jar, and screw down the cap.

6. Secure the thermometer to the jar lid by molding clay around the thermometer on the outside of the lid. The aluminum wings should be in the center of the jar.

7. Read the temperature on the thermometer. Record this as room temperature.

8. Place the jar on a windowsill in the sunlight. Turn the jar so that the black wings are angled toward the sun.

9. Watch the thermometer until the temperature reading stops rising. Record the temperature in your ScienceLog.

10. Remove the jar from direct sunlight, and allow it to return to room temperature.

Gordon Zibelman
Drexel Hill Middle School
Drexel Hill, Pennsylvania

11. Remove any shade or reflector from the lamp. Place the lamp at one end of a table.

12. Place the jar about 30 cm from the lamp. Turn the jar so that the wings are angled toward the lamp.

13. Turn on the lamp, and wait about 1 minute.

14. Move the jar a few centimeters toward the lamp until the temperature reading starts to rise. When the temperature stops rising, compare it with the reading you took in step 9.

15. Repeat step 14 until the temperature matches the temperature you recorded in step 9.

16. If the temperature reading rises too high, move the jar away from the lamp and allow it to cool. Once the reading has dropped to at least 5°C below the temperature you recorded in step 9, you may begin again at step 12.

17. When the temperature in the jar matches the temperature you recorded in step 9, record the distance between the center of the light bulb and the thermometer bulb in your ScienceLog.

Analysis

18. The thermometer measured the same amount of energy absorbed by the jar at the distance you measured to the lamp. In other words, your jar absorbed as much energy from the sun at a distance of 150 million kilometers as it did from the 100 W light bulb at the distance you recorded in step 17.

19. Use the following formula to calculate the power of the sun (be sure to show your work):

$$\frac{\text{power of the sun}}{(\text{distance to the sun})^2} = \frac{\text{power of the lamp}}{(\text{distance to the lamp})^2}$$

> **Hint:** (distance)^2 means that you multiply the distance by itself. If you found that the lamp was 5 cm away from the jar, for example, the (distance)^2 would be 25.
>
> **Hint:** Convert 150,000,000 km to 15,000,000,000,000 cm.

20. Review the discussion of scientific notation in the Math Refresher found in the Appendix at the back of this book. You will need to understand this technique for writing large numbers in order to compare your calculation with the actual figure. For practice, convert the distance to the sun given in step 19 to scientific notation.

$$15,000,000,000,000 \text{ cm} = 1.5 \times 10^{\underline{?}} \text{ cm}$$

21. The sun emits 3.7×10^{26} W of power. Compare your answer in step 19 with this value. Was this a good way to calculate the power of the sun? Explain.

 Datasheets for LabBook

 Math Skills Worksheet "What Is Scientific Notation?"

 Math Skills Worksheet "Multiplying and Dividing in Scientific Notation"

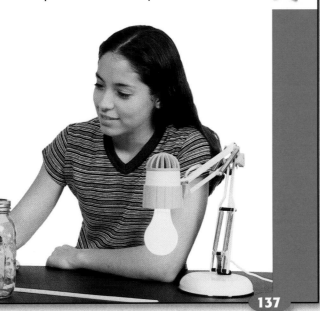

Concept Mapping: A Way to Bring Ideas Together

What Is a Concept Map?

Have you ever tried to tell someone about a book or a chapter you've just read and found that you can remember only a few isolated words and ideas? Or maybe you've memorized facts for a test and then weeks later discovered you're not even sure what topics those facts covered.

In both cases, you may have understood the ideas or concepts by themselves but not in relation to one another. If you could somehow link the ideas together, you would probably understand them better and remember them longer. This is something a concept map can help you do. A concept map is a way to see how ideas or concepts fit together. It can help you see the "big picture."

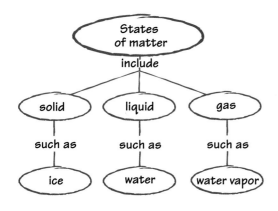

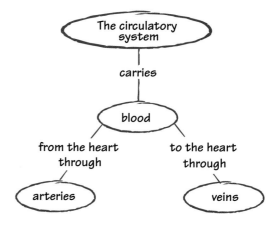

How to Make a Concept Map

❶ Make a list of the main ideas or concepts.

It might help to write each concept on its own slip of paper. This will make it easier to rearrange the concepts as many times as necessary to make sense of how the concepts are connected. After you've made a few concept maps this way, you can go directly from writing your list to actually making the map.

❷ Arrange the concepts in order from the most general to the most specific.

Put the most general concept at the top and circle it. Ask yourself, "How does this concept relate to the remaining concepts?" As you see the relationships, arrange the concepts in order from general to specific.

❸ Connect the related concepts with lines.

❹ On each line, write an action word or short phrase that shows how the concepts are related.

Look at the concept maps on this page, and then see if you can make one for the following terms:

plants, water, photosynthesis, carbon dioxide, sun's energy

One possible answer is provided at right, but don't look at it until you try the concept map yourself.

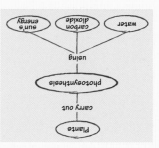

SI Measurement

The International System of Units, or SI, is the standard system of measurement used by many scientists. Using the same standards of measurement makes it easier for scientists to communicate with one another.

SI works by combining prefixes and base units. Each base unit can be used with different prefixes to define smaller and larger quantities. The table below lists common SI prefixes.

SI Prefixes

Prefix	Abbreviation	Factor	Example
kilo-	k	1,000	kilogram, 1 kg = 1,000 g
hecto-	h	100	hectoliter, 1 hL = 100 L
deka-	da	10	dekameter, 1 dam = 10 m
		1	meter, liter
deci-	d	0.1	decigram, 1 dg = 0.1 g
centi-	c	0.01	centimeter, 1 cm = 0.01 m
milli-	m	0.001	milliliter, 1 mL = 0.001 L
micro-	μ	0.000 001	micrometer, 1 μm = 0.000 001 m

SI Conversion Table

SI units	From SI to English	From English to SI
Length		
kilometer (km) = 1,000 m	1 km = 0.621 mi	1 mi = 1.609 km
meter (m) = 100 cm	1 m = 3.281 ft	1 ft = 0.305 m
centimeter (cm) = 0.01 m	1 cm = 0.394 in.	1 in. = 2.540 cm
millimeter (mm) = 0.001 m	1 mm = 0.039 in.	
micrometer (μm) = 0.000 001 m		
nanometer (nm) = 0.000 000 001 m		
Area		
square kilometer (km^2) = 100 hectares	1 km^2 = 0.386 mi^2	1 mi^2 = 2.590 km^2
hectare (ha) = 10,000 m^2	1 ha = 2.471 acres	1 acre = 0.405 ha
square meter (m^2) = 10,000 cm^2	1 m^2 = 10.765 ft^2	1 ft^2 = 0.093 m^2
square centimeter (cm^2) = 100 mm^2	1 cm^2 = 0.155 $in.^2$	1 $in.^2$ = 6.452 cm^2
Volume		
liter (L) = 1,000 mL = 1 dm^3	1 L = 1.057 fl qt	1 fl qt = 0.946 L
milliliter (mL) = 0.001 L = 1 cm^3	1 mL = 0.034 fl oz	1 fl oz = 29.575 mL
microliter (μL) = 0.000 001 L		
Mass		
kilogram (kg) = 1,000 g	1 kg = 2.205 lb	1 lb = 0.454 kg
gram (g) = 1,000 mg	1 g = 0.035 oz	1 oz = 28.349 g
milligram (mg) = 0.001 g		
microgram (μg) = 0.000 001 g		

Temperature Scales

Temperature can be expressed using three different scales: Fahrenheit, Celsius, and Kelvin. The SI unit for temperature is the kelvin (K).

Although 0 K is much colder than 0°C, a change of 1 K is equal to a change of 1°C.

Three Temperature Scales

	Fahrenheit	Celsius	Kelvin
Water boils	212°	100°	373
Body temperature	98.6°	37°	310
Room temperature	68°	20°	293
Water freezes	32°	0°	273

Temperature Conversions Table

To convert	Use this equation:	Example
Celsius to Fahrenheit °C → °F	$°F = \left(\dfrac{9}{5} \times °C\right) + 32$	Convert 45°C to °F. $°F = \left(\dfrac{9}{5} \times 45°C\right) + 32 = 113°F$
Fahrenheit to Celsius °F → °C	$°C = \dfrac{5}{9} \times (°F - 32)$	Convert 68°F to °C. $°C = \dfrac{5}{9} \times (68°F - 32) = 20°C$
Celsius to Kelvin °C → K	$K = °C + 273$	Convert 45°C to K. $K = 45°C + 273 = 318 K$
Kelvin to Celsius K → °C	$°C = K - 273$	Convert 32 K to °C. $°C = 32 K - 273 = -241°C$

Measuring Skills

Using a Graduated Cylinder

When using a graduated cylinder to measure volume, keep the following procedures in mind:

❶ Make sure the cylinder is on a flat, level surface.

❷ Move your head so that your eye is level with the surface of the liquid.

❸ Read the mark closest to the liquid level. On glass graduated cylinders, read the mark closest to the center of the curve in the liquid's surface.

Using a Meterstick or Metric Ruler

When using a meterstick or metric ruler to measure length, keep the following procedures in mind:

❶ Place the ruler firmly against the object you are measuring.

❷ Align one edge of the object exactly with the zero end of the ruler.

❸ Look at the other edge of the object to see which of the marks on the ruler is closest to that edge. **Note:** Each small slash between the centimeters represents a millimeter, which is one-tenth of a centimeter.

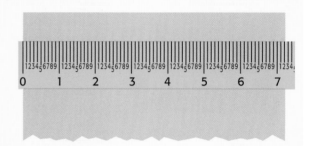

Using a Triple-Beam Balance

When using a triple-beam balance to measure mass, keep the following procedures in mind:

❶ Make sure the balance is on a level surface.

❷ Place all of the countermasses at zero. Adjust the balancing knob until the pointer rests at zero.

❸ Place the object you wish to measure on the pan. **Caution:** Do not place hot objects or chemicals directly on the balance pan.

❹ Move the largest countermass along the beam to the right until it is at the last notch that does not tip the balance. Follow the same procedure with the next-largest countermass. Then move the smallest countermass until the pointer rests at zero.

❺ Add the readings from the three beams together to determine the mass of the object.

❻ When determining the mass of crystals or powders, use a piece of filter paper. First find the mass of the paper. Then add the crystals or powder to the paper and re-measure. The actual mass of the crystals or powder is the total mass minus the mass of the paper. When finding the mass of liquids, first find the mass of the empty container. Then find the mass of the liquid and container together. The mass of the liquid is the total mass minus the mass of the container.

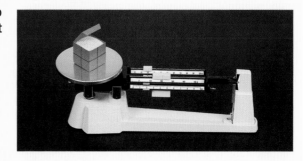

Scientific Method

The series of steps that scientists use to answer questions and solve problems is often called the **scientific method.** The scientific method is not a rigid procedure. Scientists may use all of the steps or just some of the steps of the scientific method. They may even repeat some of the steps. The goal of the scientific method is to come up with reliable answers and solutions.

Six Steps of the Scientific Method

Ask a Question

1 **Ask a Question** Good questions come from careful **observations.** You make observations by using your senses to gather information. Sometimes you may use instruments, such as microscopes and telescopes, to extend the range of your senses. As you observe the natural world, you will discover that you have many more questions than answers. These questions drive the scientific method.

Questions beginning with *what, why, how,* and *when* are very important in focusing an investigation, and they often lead to a hypothesis. (You will learn what a hypothesis is in the next step.) Here is an example of a question that could lead to further investigation.

Question: How does acid rain affect plant growth?

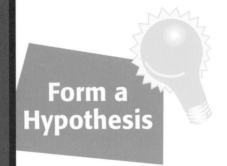

Form a Hypothesis

2 **Form a Hypothesis** After you come up with a question, you need to turn the question into a **hypothesis.** A hypothesis is a clear statement of what you expect the answer to your question to be. Your hypothesis will represent your best "educated guess" based on your observations and what you already know. A good hypothesis is testable. If observations and information cannot be gathered or if an experiment cannot be designed to test your hypothesis, it is untestable, and the investigation can go no further.

Here is a hypothesis that could be formed from the question, "How does acid rain affect plant growth?"

Hypothesis: Acid rain causes plants to grow more slowly.

Notice that the hypothesis provides some specifics that lead to methods of testing. The hypothesis can also lead to predictions. A **prediction** is what you think will be the outcome of your experiment or data collection. Predictions are usually stated in an "if . . . then" format. For example, **if** meat is kept at room temperature, **then** it will spoil faster than meat kept in the refrigerator. More than one prediction can be made for a single hypothesis. Here is a sample prediction for the hypothesis that acid rain causes plants to grow more slowly.

Prediction: If a plant is watered with only acid rain (which has a pH of 4), then the plant will grow at half its normal rate.

3 **Test the Hypothesis** After you have formed a hypothesis and made a prediction, you should test your hypothesis. There are different ways to do this. Perhaps the most familiar way is to conduct a **controlled experiment.** A controlled experiment tests only one factor at a time. A controlled experiment has a **control group** and one or more **experimental groups.** All the factors for the control and experimental groups are the same except for one factor, which is called the **variable.** By changing only one factor, you can see the results of just that one change.

Sometimes, the nature of an investigation makes a controlled experiment impossible. For example, dinosaurs have been extinct for millions of years, and the Earth's core is surrounded by thousands of meters of rock. It would be difficult, if not impossible, to conduct controlled experiments on such things. Under such circumstances, a hypothesis may be tested by making detailed observations. Taking measurements is one way of making observations.

Test the Hypothesis

4 **Analyze the Results** After you have completed your experiments, made your observations, and collected your data, you must analyze all the information you have gathered. Tables and graphs are often used in this step to organize the data.

Analyze the Results

5 **Draw Conclusions** Based on the analysis of your data, you should conclude whether or not your results support your hypothesis. If your hypothesis is supported, you (or others) might want to repeat the observations or experiments to verify your results. If your hypothesis is not supported by the data, you may have to check your procedure for errors. You may even have to reject your hypothesis and make a new one. If you cannot draw a conclusion from your results, you may have to try the investigation again or carry out further observations or experiments.

Draw Conclusions

Do they support your hypothesis?

No

Yes

6 **Communicate Results** After any scientific investigation, you should report your results. By doing a written or oral report, you let others know what you have learned. They may want to repeat your investigation to see if they get the same results. Your report may even lead to another question, which in turn may lead to another investigation.

Communicate Results

Scientific Method in Action

The scientific method is not a "straight line" of steps. It contains loops in which several steps may be repeated over and over again, while others may not be necessary. For example, sometimes scientists will find that testing one hypothesis raises new questions and new hypotheses to be tested. And sometimes, testing the hypothesis leads directly to a conclusion. Furthermore, the steps in the scientific method are not always used in the same order. Follow the steps in the diagram below, and see how many different directions the scientific method can take you.

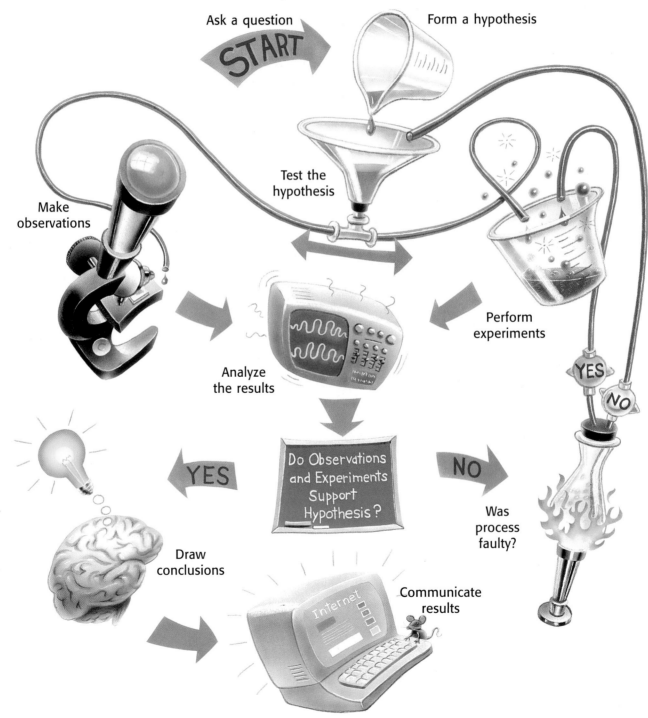

Ask a question

START

Form a hypothesis

Make observations

Test the hypothesis

Perform experiments

Analyze the results

YES

NO

YES

Do Observations and Experiments Support Hypothesis?

NO

Was process faulty?

Draw conclusions

Communicate results

Internet

Making Charts and Graphs

Circle Graphs

A circle graph, or pie chart, shows how each group of data relates to all of the data. Each part of the circle represents a category of the data. The entire circle represents all of the data. For example, a biologist studying a hardwood forest in Wisconsin found that there were five different types of trees. The data table at right summarizes the biologist's findings.

Wisconsin Hardwood Trees	
Type of tree	**Number found**
Oak	600
Maple	750
Beech	300
Birch	1,200
Hickory	150
Total	3,000

How to Make a Circle Graph

❶ In order to make a circle graph of this data, first find the percentage of each type of tree. To do this, divide the number of individual trees by the total number of trees and multiply by 100.

$$\frac{600 \text{ oak}}{3,000 \text{ trees}} \times 100 = 20\%$$

$$\frac{750 \text{ maple}}{3,000 \text{ trees}} \times 100 = 25\%$$

$$\frac{300 \text{ beech}}{3,000 \text{ trees}} \times 100 = 10\%$$

$$\frac{1,200 \text{ birch}}{3,000 \text{ trees}} \times 100 = 40\%$$

$$\frac{150 \text{ hickory}}{3,000 \text{ trees}} \times 100 = 5\%$$

❷ Now determine the size of the pie shapes that make up the chart. Do this by multiplying each percentage by 360°. Remember that a circle contains 360°.

$20\% \times 360° = 72°$ $25\% \times 360° = 90°$
$10\% \times 360° = 36°$ $40\% \times 360° = 144°$
$5\% \times 360° = 18°$

❸ Then check that the sum of the percentages is 100 and the sum of the degrees is 360.

$20\% + 25\% + 10\% + 40\% + 5\% = 100\%$
$72° + 90° + 36° + 144° + 18° = 360°$

❹ Use a compass to draw a circle and mark its center.

❺ Then use a protractor to draw angles of 72°, 90°, 36°, 144°, and 18° in the circle.

❻ Finally, label each part of the graph, and choose an appropriate title.

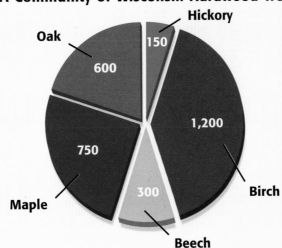

A Community of Wisconsin Hardwood Trees

Line Graphs

Line graphs are most often used to demonstrate continuous change. For example, Mr. Smith's science class analyzed the population records for their hometown, Appleton, between 1900 and 2000. Examine the data at left.

Because the year and the population change, they are the *variables*. The population is determined by, or dependent on, the year. Therefore, the population is called the **dependent variable**, and the year is called the **independent variable**. Each set of data is called a **data pair**. To prepare a line graph, data pairs must first be organized in a table like the one at left.

Population of Appleton, 1900–2000	
Year	Population
1900	1,800
1920	2,500
1940	3,200
1960	3,900
1980	4,600
2000	5,300

How to Make a Line Graph

❶ Place the independent variable along the horizontal (*x*) axis. Place the dependent variable along the vertical (*y*) axis.

❷ Label the *x*-axis "Year" and the *y*-axis "Population." Look at your largest and smallest values for the population. Determine a scale for the *y*-axis that will provide enough space to show these values. You must use the same scale for the entire length of the axis. Find an appropriate scale for the *x*-axis too.

❸ Choose reasonable starting points for each axis.

❹ Plot the data pairs as accurately as possible.

❺ Choose a title that accurately represents the data.

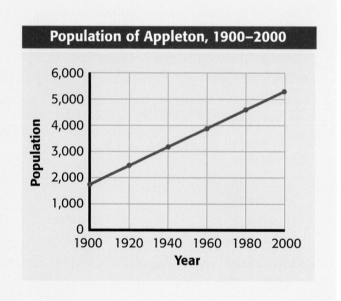

How to Determine Slope

Slope is the ratio of the change in the *y*-axis to the change in the *x*-axis, or "rise over run."

❶ Choose two points on the line graph. For example, the population of Appleton in 2000 was 5,300 people. Therefore, you can define point *a* as (2000, 5,300). In 1900, the population was 1,800 people. Define point *b* as (1900, 1,800).

❷ Find the change in the *y*-axis.
(*y* at point *a*) − (*y* at point *b*)
5,300 people − 1,800 people = 3,500 people

❸ Find the change in the *x*-axis.
(*x* at point *a*) − (*x* at point *b*)
2000 − 1900 = 100 years

❹ Calculate the slope of the graph by dividing the change in *y* by the change in *x*.

$$slope = \frac{change\ in\ y}{change\ in\ x}$$

$$slope = \frac{3,500\ people}{100\ years}$$

$$slope = 35\ people\ per\ year$$

In this example, the population in Appleton increased by a fixed amount each year. The graph of this data is a straight line. Therefore, the relationship is **linear.** When the graph of a set of data is not a straight line, the relationship is **nonlinear.**

Using Algebra to Determine Slope

The equation in step 4 may also be arranged to be:

$$y = kx$$

where y represents the change in the y-axis, k represents the slope, and x represents the change in the x-axis.

$$\text{slope} = \frac{\text{change in } y}{\text{change in } x}$$

$$k = \frac{y}{x}$$

$$k \times x = \frac{y \times x}{x}$$

$$kx = y$$

Bar Graphs

Bar graphs are used to demonstrate change that is not continuous. These graphs can be used to indicate trends when the data are taken over a long period of time. A meteorologist gathered the precipitation records at right for Hartford, Connecticut, for April 1–15, 1996, and used a bar graph to represent the data.

Precipitation in Hartford, Connecticut April 1–15, 1996

Date	Precipitation (cm)	Date	Precipitation (cm)
April 1	0.5	April 9	0.25
April 2	1.25	April 10	0.0
April 3	0.0	April 11	1.0
April 4	0.0	April 12	0.0
April 5	0.0	April 13	0.25
April 6	0.0	April 14	0.0
April 7	0.0	April 15	6.50
April 8	1.75		

How to Make a Bar Graph

❶ Use an appropriate scale and a reasonable starting point for each axis.

❷ Label the axes, and plot the data.

❸ Choose a title that accurately represents the data.

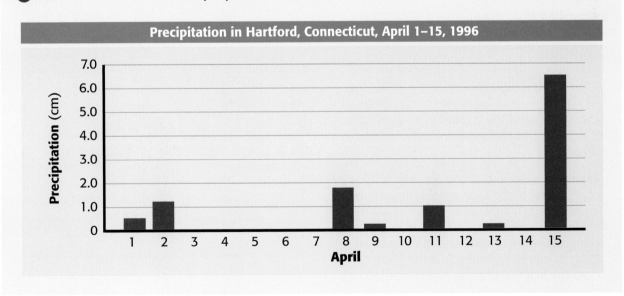

Precipitation in Hartford, Connecticut, April 1–15, 1996

Math Refresher

Science requires an understanding of many math concepts. The following pages will help you review some important math skills.

Averages

An **average,** or **mean,** simplifies a list of numbers into a single number that *approximates* their value.

> **Example:** Find the average of the following set of numbers: 5, 4, 7, and 8.

Step 1: Find the sum.

$$5 + 4 + 7 + 8 = 24$$

Step 2: Divide the sum by the amount of numbers in your set. Because there are four numbers in this example, divide the sum by 4.

$$\frac{24}{4} = 6$$

The average, or mean, is **6.**

Ratios

A **ratio** is a comparison between numbers, and it is usually written as a fraction.

> **Example:** Find the ratio of thermometers to students if you have 36 thermometers and 48 students in your class.

Step 1: Make the ratio.

$$\frac{36 \text{ thermometers}}{48 \text{ students}}$$

Step 2: Reduce the fraction to its simplest form.

$$\frac{36}{48} = \frac{36 \div 12}{48 \div 12} = \frac{3}{4}$$

The ratio of thermometers to students is **3 to 4,** or $\frac{3}{4}$. The ratio may also be written in the form 3:4.

Proportions

A **proportion** is an equation that states that two ratios are equal.

$$\frac{3}{1} = \frac{12}{4}$$

To solve a proportion, first multiply across the equal sign. This is called cross-multiplication. If you know three of the quantities in a proportion, you can use cross-multiplication to find the fourth.

> **Example:** Imagine that you are making a scale model of the solar system for your science project. The diameter of Jupiter is 11.2 times the diameter of the Earth. If you are using a plastic-foam ball with a diameter of 2 cm to represent the Earth, what diameter does the ball representing Jupiter need to be?
>
> $$\frac{11.2}{1} = \frac{x}{2 \text{ cm}}$$

Step 1: Cross-multiply.

$$\frac{11.2}{1} \times \frac{x}{2}$$

$$11.2 \times 2 = x \times 1$$

Step 2: Multiply.

$$22.4 = x \times 1$$

Step 3: Isolate the variable by dividing both sides by 1.

$$x = \frac{22.4}{1}$$
$$x = 22.4 \text{ cm}$$

You will need to use a ball with a diameter of **22.4 cm** to represent Jupiter.

Percentages

A **percentage** is a ratio of a given number to 100.

> **Example:** What is 85 percent of 40?

Step 1: Rewrite the percentage by moving the decimal point two places to the left.

$$.85$$

Step 2: Multiply the decimal by the number you are calculating the percentage of.

$$0.85 \times 40 = 34$$

85 percent of 40 is **34.**

Decimals

To **add** or **subtract decimals,** line up the digits vertically so that the decimal points line up. Then add or subtract the columns from right to left, carrying or borrowing numbers as necessary.

> **Example:** Add the following numbers: 3.1415 and 2.96.

Step 1: Line up the digits vertically so that the decimal points line up.

$$\begin{array}{r} 3.1415 \\ + 2.96 \\ \hline \end{array}$$

Step 2: Add the columns from right to left, carrying when necessary.

$$\begin{array}{r} \scriptstyle 1\ 1 \\ 3.1415 \\ + 2.96 \\ \hline 6.1015 \end{array}$$

The sum is **6.1015.**

Fractions

Numbers tell you how many; **fractions** tell you *how much of a whole.*

> **Example:** Your class has 24 plants. Your teacher instructs you to put 5 in a shady spot. What fraction does this represent?

Step 1: Write a fraction with the total number of parts in the whole as the denominator.

$$\frac{?}{24}$$

Step 2: Write the number of parts of the whole being represented as the numerator.

$$\frac{5}{24}$$

$\frac{5}{24}$ of the plants will be in the shade.

Reducing Fractions

It is usually best to express a fraction in simplest form. This is called *reducing* a fraction.

> **Example:** Reduce the fraction $\frac{30}{45}$ to its simplest form.

Step 1: Find the largest whole number that will divide evenly into both the numerator and denominator. This number is called the greatest common factor (GCF).

factors of the numerator 30: 1, 2, 3, 5, 6, 10, **15,** 30

factors of the denominator 45: 1, 3, 5, 9, **15,** 45

Step 2: Divide both the numerator and the denominator by the GCF, which in this case is 15.

$$\frac{30}{45} = \frac{30 \div 15}{45 \div 15} = \frac{2}{3}$$

$\frac{30}{45}$ reduced to its simplest form is $\frac{2}{3}$.

Adding and Subtracting Fractions

To **add** or **subtract fractions** that have the **same denominator,** simply add or subtract the numerators.

Examples:

$$\frac{3}{5} + \frac{1}{5} = ? \text{ and } \frac{3}{4} - \frac{1}{4} = ?$$

Step 1: Add or subtract the numerators.

$$\frac{3}{5} + \frac{1}{5} = \frac{4}{} \text{ and } \frac{3}{4} - \frac{1}{4} = \frac{2}{}$$

Step 2: Write the sum or difference over the denominator.

$$\frac{3}{5} + \frac{1}{5} = \frac{4}{5} \text{ and } \frac{3}{4} - \frac{1}{4} = \frac{2}{4}$$

Step 3: If necessary, reduce the fraction to its simplest form.

$$\frac{4}{5} \text{ cannot be reduced, and } \frac{2}{4} = \frac{1}{2}.$$

To **add** or **subtract fractions** that have **different denominators,** first find the least common denominator (LCD).

Examples:

$$\frac{1}{2} + \frac{1}{6} = ? \text{ and } \frac{3}{4} - \frac{2}{3} = ?$$

Step 1: Write the equivalent fractions with a common denominator.

$$\frac{3}{6} + \frac{1}{6} = ? \text{ and } \frac{9}{12} - \frac{8}{12} = ?$$

Step 2: Add or subtract.

$$\frac{3}{6} + \frac{1}{6} = \frac{4}{6} \text{ and } \frac{9}{12} - \frac{8}{12} = \frac{1}{12}$$

Step 3: If necessary, reduce the fraction to its simplest form.

$$\frac{4}{6} = \frac{2}{3}, \text{ and } \frac{1}{12} \text{ cannot be reduced.}$$

Multiplying Fractions

To **multiply fractions,** multiply the numerators and the denominators together, and then reduce the fraction to its simplest form.

Example:

$$\frac{5}{9} \times \frac{7}{10} = ?$$

Step 1: Multiply the numerators and denominators.

$$\frac{5}{9} \times \frac{7}{10} = \frac{5 \times 7}{9 \times 10} = \frac{35}{90}$$

Step 2: Reduce.

$$\frac{35}{90} = \frac{35 \div 5}{90 \div 5} = \frac{7}{18}$$

Dividing Fractions

To **divide fractions,** first rewrite the divisor (the number you divide *by*) upside down. This is called the reciprocal of the divisor. Then you can multiply and reduce if necessary.

Example:

$$\frac{5}{8} \div \frac{3}{2} = ?$$

Step 1: Rewrite the divisor as its reciprocal.

$$\frac{3}{2} \rightarrow \frac{2}{3}$$

Step 2: Multiply.

$$\frac{5}{8} \times \frac{2}{3} = \frac{5 \times 2}{8 \times 3} = \frac{10}{24}$$

Step 3: Reduce.

$$\frac{10}{24} = \frac{10 \div 2}{24 \div 2} = \frac{5}{12}$$

Scientific Notation

Scientific notation is a short way of representing very large and very small numbers without writing all of the place-holding zeros.

Example: Write 653,000,000 in scientific notation.

Step 1: Write the number without the place-holding zeros.

653

Step 2: Place the decimal point after the first digit.

6.53

Step 3: Find the exponent by counting the number of places that you moved the decimal point.

6.53000000

The decimal point was moved eight places to the left. Therefore, the exponent of 10 is positive 8. Remember, if the decimal point had moved to the right, the exponent would be negative.

Step 4: Write the number in scientific notation.

$$6.53 \times 10^8$$

Area

Area is the number of square units needed to cover the surface of an object.

Formulas:
Area of a square = side × side
Area of a rectangle = length × width
Area of a triangle = $\frac{1}{2}$ × base × height

Examples: Find the areas.

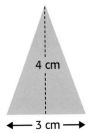

Triangle
Area = $\frac{1}{2}$ × base × height

Area = $\frac{1}{2}$ × 3 cm × 4 cm

Area = **6 cm²**

4 cm

3 cm

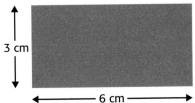

Rectangle
Area = length × width
Area = 6 cm × 3 cm
Area = **18 cm²**

3 cm

6 cm

3 cm

Square
Area = side × side
Area = 3 cm × 3 cm
Area = **9 cm²**

3 cm

Volume

Volume is the amount of space something occupies.

Formulas:
Volume of a cube =
side × side × side

Volume of a prism =
area of base × height

Examples:
Find the volume
of the solids.

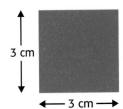

Cube
Volume = side × side × side
Volume = 4 cm × 4 cm × 4 cm
Volume = **64 cm³**

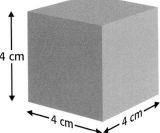

4 cm

4 cm

4 cm

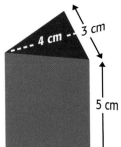

4 cm

3 cm

5 cm

Prism
Volume = area of base × height
Volume = (area of triangle) × height
Volume = $\left(\frac{1}{2} \times 3 \text{ cm} \times 4 \text{ cm}\right) \times 5$ cm
Volume = 6 cm² × 5 cm
Volume = **30 cm³**

Periodic Table of the Elements

Each square on the table includes an element's name, chemical symbol, atomic number, and atomic mass.

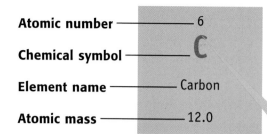

Atomic number	6
Chemical symbol	C
Element name	Carbon
Atomic mass	12.0

The background color indicates the type of element. Carbon is a nonmetal.

The color of the chemical symbol indicates the physical state at room temperature. Carbon is a solid.

Background

Metals ▮

Metalloids ▮

Nonmetals ▮

Chemical Symbol

Solid ▮

Liquid ▮

Gas ▮

Period 1

| 1 H Hydrogen 1.0 |

	Group 1	Group 2
Period 2	3 Li Lithium 6.9	4 Be Beryllium 9.0
Period 3	11 Na Sodium 23.0	12 Mg Magnesium 24.3

	Group 1	Group 2	Group 3	Group 4	Group 5	Group 6	Group 7	Group 8	Group 9
Period 4	19 K Potassium 39.1	20 Ca Calcium 40.1	21 Sc Scandium 45.0	22 Ti Titanium 47.9	23 V Vanadium 50.9	24 Cr Chromium 52.0	25 Mn Manganese 54.9	26 Fe Iron 55.8	27 Co Cobalt 58.9
Period 5	37 Rb Rubidium 85.5	38 Sr Strontium 87.6	39 Y Yttrium 88.9	40 Zr Zirconium 91.2	41 Nb Niobium 92.9	42 Mo Molybdenum 95.9	43 Tc Technetium (97.9)	44 Ru Ruthenium 101.1	45 Rh Rhodium 102.9
Period 6	55 Cs Cesium 132.9	56 Ba Barium 137.3	57 La Lanthanum 138.9	72 Hf Hafnium 178.5	73 Ta Tantalum 180.9	74 W Tungsten 183.8	75 Re Rhenium 186.2	76 Os Osmium 190.2	77 Ir Iridium 192.2
Period 7	87 Fr Francium (223.0)	88 Ra Radium (226.0)	89 Ac Actinium (227.0)	104 Rf Rutherfordium (261.1)	105 Db Dubnium (262.1)	106 Sg Seaborgium (263.1)	107 Bh Bohrium (262.1)	108 Hs Hassium (265)	109 Mt Meitnerium (266)

A row of elements is called a period.

A column of elements is called a group or family.

Lanthanides

| 58 Ce Cerium 140.1 | 59 Pr Praseodymium 140.9 | 60 Nd Neodymium 144.2 | 61 Pm Promethium (144.9) | 62 Sm Samarium 150.4 |

Actinides

| 90 Th Thorium 232.0 | 91 Pa Protactinium 231.0 | 92 U Uranium 238.0 | 93 Np Neptunium (237.0) | 94 Pu Plutonium 244.1 |

These elements are placed below the table to allow the table to be narrower.

This zigzag line reminds you where the metals, nonmetals, and metalloids are.

A number in parenthesis is the mass number of the most stable form of that element.

Group 18

			Group 13	Group 14	Group 15	Group 16	Group 17	2 **He** Helium 4.0
			5 **B** Boron 10.8	6 **C** Carbon 12.0	7 **N** Nitrogen 14.0	8 **O** Oxygen 16.0	9 **F** Fluorine 19.0	10 **Ne** Neon 20.2
Group 10	**Group 11**	**Group 12**	13 **Al** Aluminum 27.0	14 **Si** Silicon 28.1	15 **P** Phosphorus 31.0	16 **S** Sulfur 32.1	17 **Cl** Chlorine 35.5	18 **Ar** Argon 39.9
28 **Ni** Nickel 58.7	29 **Cu** Copper 63.5	30 **Zn** Zinc 65.4	31 **Ga** Gallium 69.7	32 **Ge** Germanium 72.6	33 **As** Arsenic 74.9	34 **Se** Selenium 79.0	35 **Br** Bromine 79.9	36 **Kr** Krypton 83.8
46 **Pd** Palladium 106.4	47 **Ag** Silver 107.9	48 **Cd** Cadmium 112.4	49 **In** Indium 114.8	50 **Sn** Tin 118.7	51 **Sb** Antimony 121.8	52 **Te** Tellurium 127.6	53 **I** Iodine 126.9	54 **Xe** Xenon 131.3
78 **Pt** Platinum 195.1	79 **Au** Gold 197.0	80 **Hg** Mercury 200.6	81 **Tl** Thallium 204.4	82 **Pb** Lead 207.2	83 **Bi** Bismuth 209.0	84 **Po** Polonium (209.0)	85 **At** Astatine (210.0)	86 **Rn** Radon (222.0)
110 **Uun*** Ununnilium (271)	111 **Uuu*** Unununium (272)	112 **Uub*** Ununbium (277)		114 **Uuq*** Ununquadium (285)		116 **Uuh*** Ununhexium (289)		118 **Uuo*** Ununoctium (293)

63 **Eu** Europium 152.0	64 **Gd** Gadolinium 157.3	65 **Tb** Terbium 158.9	66 **Dy** Dysprosium 162.5	67 **Ho** Holmium 164.9	68 **Er** Erbium 167.3	69 **Tm** Thulium 168.9	70 **Yb** Ytterbium 173.0	71 **Lu** Lutetium 175.0
95 **Am** Americium (243.1)	96 **Cm** Curium (247.1)	97 **Bk** Berkelium (247.1)	98 **Cf** Californium (251.1)	99 **Es** Einsteinium (252.1)	100 **Fm** Fermium (257.1)	101 **Md** Mendelevium (258.1)	102 **No** Nobelium (259.1)	103 **Lr** Lawrencium (262.1)

The official names and symbols for the elements greater than 109 will eventually be approved by a committee of scientists.

Physical Science Refresher

Atoms and Elements

Every object in the universe is made up of particles of some kind of matter. **Matter** is anything that takes up space and has mass. All matter is made up of elements. An **element** is a substance that cannot be separated into simpler components by ordinary chemical means. This is because each element consists of only one kind of atom. An **atom** is the smallest unit of an element that has all of the properties of that element.

Atomic Structure

Atoms are made up of small particles called subatomic particles. The three major types of subatomic particles are **electrons, protons,** and **neutrons.** Electrons have a negative electric charge, protons have a positive charge, and neutrons have no electric charge. The protons and neutrons are packed close to one another to form the **nucleus.** The protons give the nucleus a positive charge. Electrons are most likely to be found in regions around the nucleus called **electron clouds.** The negatively charged electrons are attracted to the positively charged nucleus. An atom may have several energy levels in which electrons are located.

Atomic Number

To help in the identification of elements, scientists have assigned an **atomic number** to each kind of atom. The atomic number is the number of protons in the atom. Atoms with the same number of protons are all the same kind of element. In an uncharged, or electrically neutral, atom there are an equal number of protons and electrons. Therefore, the atomic number equals the number of electrons in an uncharged atom. The number of neutrons, however, can vary for a given element. Atoms of the same element that have different numbers of neutrons are called **isotopes.**

Periodic Table of the Elements

In the periodic table, the elements are arranged from left to right in order of increasing atomic number. Each element in the table is in a separate box. An atom of each element has one more electron and one more proton than an atom of the element to its left. Each horizontal row of the table is called a **period.** Changes in chemical properties of elements across a period correspond to changes in the electron arrangements of their atoms. Each vertical column of the table, known as a **group,** lists elements with similar properties. The elements in a group have similar chemical properties because their atoms have the same number of electrons in their outer energy level. For example, the elements helium, neon, argon, krypton, xenon, and radon all have similar properties and are known as the noble gases.

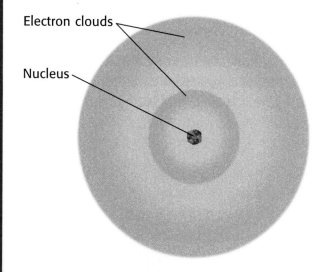

Electron clouds

Nucleus

Molecules and Compounds

When two or more elements are joined chemically, the resulting substance is called a **compound.** A compound is a new substance with properties different from those of the elements that compose it. For example, water, H_2O, is a compound formed when hydrogen (H) and oxygen (O) combine. The smallest complete unit of a compound that has the properties of that compound is called a **molecule.** A chemical formula indicates the elements in a compound. It also indicates the relative number of atoms of each element present. The chemical formula for water is H_2O, which indicates that each water molecule consists of two atoms of hydrogen and one atom of oxygen. The subscript number is used after the symbol for an element to indicate how many atoms of that element are in a single molecule of the compound.

Acids, Bases, and pH

An ion is an atom or group of atoms that has an electric charge because it has lost or gained one or more electrons. When an acid, such as hydrochloric acid, HCl, is mixed with water, it separates into ions. An **acid** is a compound that produces hydrogen ions, H^+, in water. The hydrogen ions then combine with a water molecule to form a hydronium ion, H_3O^+. A **base,** on the other hand, is a substance that produces hydroxide ions, OH^-, in water.

To determine whether a solution is acidic or basic, scientists use pH. The **pH** is a measure of the hydronium ion concentration in a solution. The pH scale ranges from 0 to 14. The middle point, pH = 7, is neutral, neither acidic nor basic. Acids have a pH less than 7; bases have a pH greater than 7. The lower the number is, the more acidic the solution. The higher the number is, the more basic the solution.

Chemical Equations

A chemical reaction occurs when a chemical change takes place. (In a chemical change, new substances with new properties are formed.) A chemical equation is a useful way of describing a chemical reaction by means of chemical formulas. The equation indicates what substances react and what the products are. For example, when carbon and oxygen combine, they can form carbon dioxide. The equation for the reaction is as follows: $C + O_2 \rightarrow CO_2$.

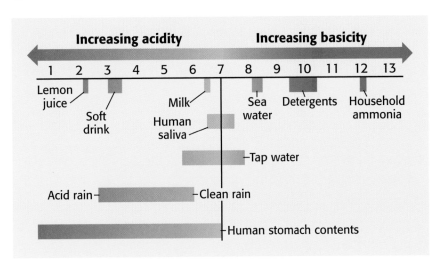

The Six Kingdoms

Kingdom Archaebacteria

The organisms in this kingdom are single-celled prokaryotes.

Archaebacteria		
Group	**Examples**	**Characteristics**
Methanogens	*Methanococcus*	found in soil, swamps, the digestive tract of mammals; produce methane gas; can't live in oxygen
Thermophiles	*Sulpholobus*	found in extremely hot environments; require sulphur, can't live in oxygen
Halophiles	*Halococcus*	found in environments with very high salt content, such as the Dead Sea; nearly all can live in oxygen

Kingdom Eubacteria

There are more than 4,000 named species in this kingdom of single-celled prokaryotes.

Eubacteria		
Group	**Examples**	**Characteristics**
Bacilli	*Escherichia coli*	rod-shaped; free-living, symbiotic, or parasitic; some can fix nitrogen; some cause disease
Cocci	*Streptococcus*	spherical-shaped, disease-causing; can form spores to resist unfavorable environments
Spirilla	*Treponema*	spiral-shaped; responsible for several serious illnesses, such as syphilis and Lyme disease

Kingdom Protista

The organisms in this kingdom are eukaryotes. There are single-celled and multicellular representatives.

Protists		
Group	**Examples**	**Characteristics**
Sacodines	*Amoeba*	radiolarians; single-celled consumers
Ciliates	*Paramecium*	single-celled consumers
Flagellates	*Trypanosoma*	single-celled parasites
Sporozoans	*Plasmodium*	single-celled parasites
Euglenas	*Euglena*	single-celled; photosynthesize
Diatoms	*Pinnularia*	most are single-celled; photosynthesize
Dinoflagellates	*Gymnodinium*	single-celled; some photosynthesize
Algae	*Volvox*, coral algae	4 phyla; single- or many-celled; photosynthesize
Slime molds	*Physarum*	single- or many-celled; consumers or decomposers
Water molds	powdery mildew	single- or many-celled, parasites or decomposers

Kingdom Fungi

There are single-celled and multicellular eukaryotes in this kingdom. There are four major groups of fungi.

Fungi		
Group	**Examples**	**Characteristics**
Threadlike fungi	bread mold	spherical; decomposers
Sac fungi	yeast, morels	saclike; parasites and decomposers
Club fungi	mushrooms, rusts, smuts	club-shaped; parasites and decomposers
Lichens	British soldier	symbiotic with algae

Kingdom Plantae

The organisms in this kingdom are multicellular eukaryotes. They have specialized organ systems for different life processes. They are classified in divisions instead of phyla.

Plants		
Group	**Examples**	**Characteristics**
Bryophytes	mosses, liverworts	reproduce by spores
Club mosses	*Lycopodium,* ground pine	reproduce by spores
Horsetails	rushes	reproduce by spores
Ferns	spleenworts, sensitive fern	reproduce by spores
Conifers	pines, spruces, firs	reproduce by seeds; cones
Cycads	*Zamia*	reproduce by seeds
Gnetophytes	*Welwitschia*	reproduce by seeds
Ginkgoes	*Ginkgo*	reproduce by seeds
Angiosperms	all flowering plants	reproduce by seeds; flowers

Kingdom Animalia

This kingdom contains multicellular eukaryotes. They have specialized tissues and complex organ systems.

Animals		
Group	**Examples**	**Characteristics**
Sponges	glass sponges	no symmetry or segmentation; aquatic
Cnidarians	jellyfish, coral	radial symmetry; aquatic
Flatworms	planaria, tapeworms, flukes	bilateral symmetry; organ systems
Roundworms	*Trichina,* hookworms	bilateral symmetry; organ systems
Annelids	earthworms, leeches	bilateral symmetry; organ systems
Mollusks	snails, octopuses	bilateral symmetry; organ systems
Echinoderms	sea stars, sand dollars	radial symmetry; organ systems
Arthropods	insects, spiders, lobsters	bilateral symmetry; organ systems
Chordates	fish, amphibians, reptiles, birds, mammals	bilateral symmetry; complex organ systems

Using the Microscope

Parts of the Compound Light Microscope

- The **ocular lens** magnifies the image 10×.

- The **low-power objective** magnifies the image 10×.

- The **high-power objective** magnifies the image either 40× or 43×.

- The **revolving nosepiece** holds the objectives and can be turned to change from one magnification to the other.

- The **body tube** maintains the correct distance between the ocular lens and objectives.

- The **coarse-adjustment knob** moves the body tube up and down to allow focusing of the image.

- The **fine-adjustment knob** moves the body tube slightly to bring the image into sharper focus.

- The **stage** supports a slide.

- **Stage clips** hold the slide in place for viewing.

- The **diaphragm** controls the amount of light coming through the stage.

- The light source provides a **light** for viewing the slide.

- The **arm** supports the body tube.

- The **base** supports the microscope.

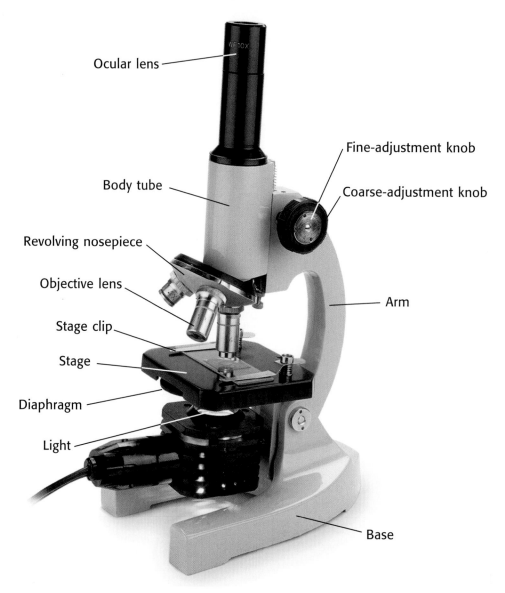

Ocular lens

Fine-adjustment knob

Coarse-adjustment knob

Body tube

Revolving nosepiece

Objective lens

Arm

Stage clip

Stage

Diaphragm

Light

Base

Proper Use of the Compound Light Microscope

1 Carry the microscope to your lab table using both hands. Place one hand beneath the base, and use the other hand to hold the arm of the microscope. Hold the microscope close to your body while moving it to your lab table.

2 Place the microscope on the lab table at least 5 cm from the edge of the table.

3 Check to see what type of light source is used by your microscope. If the microscope has a lamp, plug it in, making sure that the cord is out of the way. If the microscope has a mirror, adjust it to reflect light through the hole in the stage.
Caution: If your microscope has a mirror, do not use direct sunlight as a light source. Direct sunlight can damage your eyes.

4 Always begin work with the low-power objective in line with the body tube. Adjust the revolving nosepiece.

5 Place a prepared slide over the hole in the stage. Secure the slide with the stage clips.

6 Look through the ocular lens. Move the diaphragm to adjust the amount of light coming through the stage.

7 Look at the stage from eye level. Slowly turn the coarse adjustment to lower the objective until it almost touches the slide. Do not allow the objective to touch the slide.

8 Look through the ocular lens. Turn the coarse adjustment to raise the low-power objective until the image is in focus. Always focus by raising the objective away from the slide. *Never focus the objective downward.* Use the fine adjustment to sharpen the focus. Keep both eyes open while viewing a slide.

9 Make sure that the image is exactly in the center of your field of vision. Then switch to the high-power objective. Focus the image, using only the fine adjustment. *Never use the coarse adjustment at high power.*

10 When you are finished using the microscope, remove the slide. Clean the ocular lens and objective lenses with lens paper. Return the microscope to its storage area. Remember, you should use both hands to carry the microscope.

Making a Wet Mount

1 Use lens paper to clean a glass slide and a coverslip.

2 Place the specimen you wish to observe in the center of the slide.

3 Using a medicine dropper, place one drop of water on the specimen.

4 Hold the coverslip at the edge of the water and at a 45° angle to the slide. Make sure that the water runs along the edge of the coverslip.

5 Lower the coverslip slowly to avoid trapping air bubbles.

6 Water might evaporate from the slide as you work. Add more water to keep the specimen fresh. Place the tip of the medicine dropper next to the edge of the coverslip. Add a drop of water. (You can also use this method to add stain or solutions to a wet mount.) Remove excess water from the slide by using the corner of a paper towel as a blotter. Do not lift the coverslip to add or remove water.

Glossary

A

abiotic describes nonliving factors in the environment (4, 48)

acid precipitation precipitation that contains acids due to air pollution (106)

adaptation a characteristic that helps an organism survive in its environment (16)

algae (AL JEE) protists that convert the sun's energy into food through photosynthesis (6, 17)

alien an organism that makes a home for itself in a new place (78)

Animalia the classification kingdom containing complex, multicellular organisms that lack cell walls, are usually able to move around, and possess nervous systems that help them be aware of and react to their surroundings (157)

atmosphere a mixture of gases that surrounds a planet, such as Earth (74, 76)

atom the smallest part of an element that has all of the properties of that element (154)

B

biodegradable capable of being broken down by the environment (80)

biodiversity the number and variety of living things (79)

biome a large region characterized by a specific type of climate and certain types of plant and animal communities (48)

biosphere the part of the Earth where life exists (7)

biotic describes living factors in the environment (4)

C

carbon cycle the movement of carbon from the nonliving environment into living things and then back into the nonliving environment (31)

carnivore a consumer that eats animals (9)

carrying capacity the largest population that a given environment can support over a long period of time (15)

coal a solid fossil fuel formed underground from buried, decomposed plant material (102)

coevolution (KOH ev uh LOO shuhn) the long-term changes that take place in two species because of their close interactions with one another (18)

community all of the populations of different species that live and interact in an area (6)

Competition

competition two or more species or individuals trying to use the same limited resource (15)

compound light microscope a microscope that consists of a tube with lenses, a stage, and a light source (158)

conifer a tree that produces seeds in cones (50)

conservation the wise use of and preservation of natural resources (81)

consumer an organism that eats producers or other organisms for energy (9)

D

deciduous describes trees with leaves that change color in autumn and fall off in winter (49)

decomposer an organism that gets energy by breaking down the remains of dead organisms or animal wastes and consuming or absorbing the nutrients (9)

decomposition the breakdown of dead materials into carbon dioxide and water (32)

deep-water zone the zone of a lake or pond below the open-water zone where no light reaches (61)

deforestation the clearing of forest lands (79)

desert a hot, dry biome inhabited by organisms adapted to survive high daytime temperatures and long periods without rain (53)

diversity a measure of the number of species an area contains (51)

E

energy pyramid a diagram shaped like a triangle that shows the loss of energy at each level of the food chain (11)

energy resource a natural resource that humans use to produce energy (101)

estuary an area where fresh water from streams and rivers spills into the ocean (59)

evaporation the change of state from liquid to vapor (30)

extinct describes a species of organism that has died out completely (85)

F

food chain a diagram that represents how the energy in food molecules flows from one organism to the next (10)

food web a complex diagram representing the many energy pathways in a real ecosystem (10)

fossil fuel a nonrenewable energy resource that forms in the Earth's crust over millions of years from the buried remains of once-living organisms (101)

G

gasohol a mixture of gasoline and alcohol that is burned as a fuel (114)

geothermal energy energy from within the Earth (115)

global warming a rise in average global temperatures (31, 76)

ground water water stored in underground caverns or porous rock (31)

H

habitat the environment where an organism lives (12)

herbivore a consumer that eats plants (9)

host an organism on which a parasite lives (18)

hydroelectric energy electricity produced by falling water (113)

hypothesis a possible explanation or answer to a question (142–144)

K

kingdom the most general of the seven levels of classification (156–157)

L

limiting factor a needed resource that is in limited supply (14)

littoral zone the zone of a lake or pond closest to the edge of the land (61)

M

marine describes an ecosystem based on salty water (55)

marsh a treeless wetland ecosystem where such plants as cattails and rushes grow (62)

matter anything that occupies space and has mass (30)

meter the basic unit of length in the SI system (139)

mutualism (MYOO choo uhl IZ uhm) a symbiotic relationship in which both organisms benefit (17)

N

natural gas a gaseous fossil fuel (102)

natural resource any natural substance, organism, or energy form that living things use (98)

niche an organism's way of life and its relationships with its abiotic and biotic environments (12)

nitrogen cycle the movement of nitrogen from the nonliving environment into living organisms and back again (32)

nitrogen fixation the process of changing nitrogen gas into forms that plants can use (33)

nonrenewable resource a natural resource that cannot be replaced or that can be replaced only over thousands or millions of years (77, 82, 99-100)

nuclear energy the form of energy associated with changes in the nucleus of an atom; an alternative energy resource (108)

nuclear fusion the process by which two or more nuclei with small masses join together, or fuse, to form a larger, more massive nucleus, along with the production of energy (109)

O

observation any use of the senses to gather information (144)

omnivore a consumer that eats a variety of organisms (9)

open-water zone the zone of a lake or pond that extends from the littoral zone out across the top of the water and that is only as deep as light can reach through the water (61)

overpopulation a condition that occurs when the number of individuals within an environment becomes so large that there are not enough resources to support them all (78)

ozone a gas molecule that is made up of three oxygen atoms; absorbs ultraviolet radiation from the sun (75–76)

P

parasite an organism that feeds on another living creature, usually without killing it (18)

parasitism (PAR uh SIET IZ uhm) a symbiotic association in which one organism benefits while the other is harmed (18)

permafrost the permanently frozen ground below the soil surface in the arctic tundra (54)

petroleum an oily mixture of flammable organic compounds from which liquid fossil fuels and other products are separated; crude oil (101)

phytoplankton (FITE oh PLANK tuhn) a microscopic photosynthetic organism that floats near the surface of the ocean (55)

pioneer species the first organisms to grow in an area undergoing ecological succession; usually lichens in primary succession and fast-growing, weedy plants in secondary succession (35)

plankton   very small organisms floating at or near the ocean's surface that form the base of the ocean's food web (55)

Plantae   the kingdom that contains plants—complex, multicellular organisms that are usually green and use the sun's energy to make sugar by photosynthesis (157)

pollutant   a harmful substance in an environment (74)

pollution   the presence of harmful substances in an environment (74)

population   a group of individuals of the same species that live together in the same area at the same time (6)

precipitation   water that moves from the atmosphere to the land and ocean, including rain, snow, sleet, and hail (30)

predator   an organism that eats other organisms (16)

prey   an organism that is eaten by another organism (16)

producer   organisms that make their own food, usually by using the energy from sunlight to make sugar (8)

Protista   a kingdom of eukaryotic single-celled or simple, multicellular organisms; kingdom Protista contains all eukaryotes that are not plants, animals, or fungi (156)

R

radiation   energy transferred as waves on particles (108)

recycling   the process of making new products from reprocessed used products (83, 100)

renewable resource   a natural resource that can be used and replaced over a relatively short time period (77, 99)

resource recovery   the process of transforming into usable products things normally thrown away (84)

S

savanna   a tropical grassland biome with scattered clumps of trees (52)

scavenger   an animal that feeds on the bodies of dead animals (9)

sediment   fine particles of sand, dust, or mud that are deposited over time by wind or water (103)

smog   a photochemical fog produced by the reaction of sunlight and air pollutants (107)

soil   a loose mixture of small mineral fragments and organic material (35)

solar energy   energy from the sun (109)

strip mining   a process in which rock and soil are stripped from the Earth's surface to expose the underlying materials to be mined (105)

succession   the gradual regrowth or development of a community of organisms over time (34)

swamp   a wetland ecosystem in which trees and vines grow (63)

symbiosis   (SIM bie OH sis) a close, long-term association between two or more species (17)

T

temperature   a measure of how hot or cold something is (140)

toxic   poisonous (75)

tributary   a small stream or river that flows into a larger one (60)

tundra   a far-northern biome characterized by long, cold winters, permafrost, and few trees (54)

W

water cycle   the movement of water between the ocean, atmosphere, land, and living things (30)

wetland   an area of land where the water level is near or above the surface of the ground for most of the year (62)

wind energy   energy in wind (112)

Z

zooplankton   (ZOH oh PLANGK tuhn) protozoa that, along with the phytoplankton they consume, form the base of the ocean's food web (55)

Index

Boldface numbers refer to an illustration on that page.

Credits

Abbreviations used: (t) top, (c) center, (b) bottom, (l) left, (r) right, (bkgd) background

ILLUSTRATIONS

All illustrations, unless otherwise noted below by Holt, Rinehart and Winston.

Table of Contents Page iv(cl) Will Nelson/Sweet Reps

Scope and Sequence: T11, Paul DiMare, T13, Dan Stuckenschneider/Uhl Studios, Inc.

Chapter One Page 4 (b); 5 (l), Will Nelson/Sweet Reps; 6-7 (b) John White/The Neis Group; 8-9 (b) Will Nelson/Sweet Reps; 10 (b), John White/The Neis Group; 11 (b), Will Nelson/Sweet Reps, 13 (br), Will Nelson/Sweet Reps; 14 (bl), Blake Thornton/Rita Marie; 19 (cr), Mike Wepplo/Das Group; 22 (cl), Will Nelson/Sweet Reps; 25 (cr), Jared Schneidman Design.

Chapter Two Page 30 (b), Robert Hynes/Mendola Artists; 31 (b), Robert Hynes/Mendola Artists; 32 (b), Robert Hynes/Mendola Artists; 35, Robert Hynes/Mendola Artists; 36, Robert Hynes/Mendola Artists; 40 (br), Robert Hynes/Mendola Artists; 42 (br), Robert Hynes/Mendola Artists; 45 (cl), MapQuest.com.

Chapter Three Page 48 (b), MapQuest.com; 49 (b), Will Nelson/Sweet Reps; 50 (b), Will Nelson/Sweet Reps; 51 (b), Will Nelson/Sweet Reps; 53 (b), Will Nelson/Sweet Reps; 56-57, Yuan Lee; 60 (bl), Will Nelson/Sweet Reps; 61 (br), Mark Heine; 64 (b), Carlyn Iverson/67 (cr), Will Nelson/Sweet Reps; 68 (bl), Will Nelson/Sweet Reps; 69 (tr,cr), Rob Schuster/Hankins and Tegenborg.

Chapter Four Page 75(tr), Peter Darro; 93 (t), John White/The Neis Group.

Chapter Five: Page 98(b), Uhl Studios, Inc.; 103(bl), Uhl Studios, Inc.; 104(l), Uhl Studios, Inc.; 105(t), MapQuest.com; 108(cl), Stephen Durke/Washington Artists; 111(tr), John Huxtable/Black Creative; 115(br), Uhl Studios, Inc.; 118(br), John Huxtable/Black Creative; 118(cr), Uhl Studios, Inc.; 121(cr), Sidney Jablonski.

LabBook Page 132 (r), John White/The Neis Group.

Appendix Page 140 (t), Terry Guyer; 144 (b), Mark Mille/Sharon Langley; 152-153, Kristy Sprott, 154 (bl), Stephen Durke/Washington Artists; 155 (tl,c) Stephen Durke/Washington Artists; 155 (b), Bruce Burdick; 155 (cl), Stephen Durke/Washington Artists.

PHOTOGRAPHY

Cover and Title page: Secret Sea Visions/Peter Arnold, Inc.

Feature Borders: Unless otherwise noted below, all images ©2001 PhotoDisc/HRW. "Across the Sciences" 70, all images by HRW; "Careers" 71, 94, sand bkgd and Saturn, Corbis; DNA, Morgan Cain & Associates; scuba gear, ©1997 Radlund & Associates for Artville; "Eureka" 123, ©2001 PhotoDisc/HRW; "Eye on the Environment" 27, 122, clouds and sea in bkgd, HRW; bkgd grass, red eyed frog, Corbis ; hawks, pelican, Animals Animals/Earth Scenes; rat, Visuals Unlimited/John Grelach; endangered flower, Dan Suzio/Photo Researchers, Inc.; "Health Watch" 26, dumbbell, Sam Dudgeon/HRW Photo; aloe vera, EKG, Victoria Smith/HRW Photo; basketball, ©1997 Radlund & Associates for Artville; shoes, bubbles, Greg Geisler; "Scientific Debate" 95, Sam Dudgeon/HRW Photo; "Weird Science" 44, mite, David Burder/Stone; atom balls, J/B Woolsey Associates; walking stick, turtle, EclectiCollection.

Table of Contents: iv(tl), Sanford D. Porter/U.S. Department of Agriculture; iv(cl), Jeff Hunter/Image Bank; iv(b), Sylvian Coffie/Stone; v(tr), Owen Garrett/Centre for Atmospheric Science at Cambridge University, UK/NASA; v(cr), J. Contreras Chacel/International Stock; v(b), Peter Van Steen/HRW Photo' vi(tl), SuperStock; vi(cl), Peter Van Steen/HRW Photo; vi(b), Peter Van Steen/HRW Photo; vii(tr), E. R. Degginger/Color-Pic, Inc.; vii(cr), Lincoln P. Brower; vii(br), Gay Bumgarner/Stone.

Scope and Sequence: T8(l), Lee F. Snyder/Photo Researchers, Inc.; T8(r), Stephen Dalton/Photo Researchers, Inc.; T10, E. R. Degginger/Color-Pic, Inc., T12(l), Rob Matheson/The Stock Market

Master Materials List: T26(bl, br), Image ©2001 PhotoDisc; T27(t,bl, bc), Image ©2001 PhotoDisc

Chapter One: pp. 2-3 Hans Reinhard/Bruce Coleman, Inc.; 3 HRW Photo; 12(bl), Gamma-Liaison; 12(tl), Laguna Photo/Liaison International; 13(cr), Jeff Lepore/Photo Researchers, Inc.; 14(c), Jeff Foott/AUSCAPE; 15 Ross Hamilton/Stone; 16(tl), Visuals Unlimited/Gerald & Buff Corsi; 16(bl), Hans Pfletschinger/Peter Arnold; 17(b), Ed Robinson/Tom Stack & Associates; 17(cr), Telegraph Color Library/FPG; 17(tr), Peter Parks/Animals Animals Earth Scenes; 18(tl), Gay Bumgarner/Stone; 18(bl), Carol Hughes/Bruce Coleman; 19(tr), CSIRO Wildlife & Ecology; 21 Victoria Smith/HRW Photo; 23(cr), Gay Bumgarner/Stone; 26(tc), Darlyne Murawski/National Geographic Society Image Collection; 27(tr), Sanford D. Porter/U.S. Department of Agriculture

Chapter Two: pp 28-29 John D. Dawson/USPS; 29 HRW Photo; 33 Ray Pfortner/Peter Arnold, Inc.; 34(t), Diana L. Stratton/Tom Stack; 34(b), Stan Osolinski/FPG; 37 Kim Heacox/DRK Photo; 39(tr) Kenneth Gabrielson/Liaison International; 39(tl) Doug Sokell/Visuals Unlimited; 39(b) Larry Nielsen/Peter Arnold 40(tl), Ray Pfortner/Peter Arnold, Inc.; 41 Kim Heacox/DRK Photo; 44 Charles O'Rear/Westlight

Chapter Three: pp. 46-47 VCG/FPG International; 47 HRW Photo; 52(c), Grant Heilman; 52(b), Tom Brakefield/Bruce Coleman; 54 Kathy Bushue/Stone; 55(bc), Stuart Westmorland/Stone; 55(bl) Manfred Kage/Peter Arnold; 58(tl), Jeff Hunter/Image Bank; 58(bl), Zig Leszczynski/Animals Animals; 59(tr) Johnny Johnson/DRK Photo; 59(cr), Nancy Sefton/Photo Researchers, Inc.; 62(b), Dwight Kuhn; 62(tr), Unicorn Stock Photos; 63(cr), Don & Pat Valenti/DRK Photo; 63(tr), Hardie Truesdale; 67(tl), Jeff Hunter/Image Bank; 70 Dr. Verena Tunnicliffe; 71(tl, br), Lincoln P. Brower

Chapter Four: pp. 72-73 Peter Cade/Stone; 73 HRW Photo; 74(c), Grant Heilman; 74(bl), Arthur Tilley/Stone; 75(br), Ken Griffiths/Stone; 76(c), Roy Morsch/Stock Market; 76(tl), Owen Garrett/Centre for Atmospheric Science at Cambridge University, UK/NASA; 77 Jacques Jangoux/Stone; 78(tl), Runk/ Schoenberger/Grant Heilman; 78(cl), John Eastcott/VVA Momatiuk/Woodfin Camp; 79(tr), Rex Ziak/Stone; 79(br), Martin Rogers/Uniphoto; 80(cl), Fed Bavendam/Peter Arnold; 82(tl), Argonne National Laboratory; 82(bl), Emile Luider/Rapho/Liaison; 83(tr), PhotoEdit; 83(bl), Kay Park-Rec Co; 83(bc), J. Contreras Chacel/International Stock; 84(cl), Martin Bond/Science Photo Library/Photo Researchers, Inc.; 85(cl), Uniphoto; 85(br), K. W. Fink/Bruce Coleman; 85(tr), Toyohiro Yamada/FPG; 86 Stephen J. Krasemann/DRK Photo; 87(tr), Will & Deni McIntyre/Stone; 87(cr), Stephen J. Krasemann/DRK Photo; 90(cl), Arthur Tilley/Stone; 90(cr), K. W. Fink/Bruce Coleman, Inc.; 92 Runk/Schoenberger/ Grant Heilman; 94(tl, br), Karen M. Allen; 95(bc), Art Wolfe

Chapter Five: pp. 96-97 Novovitch/Liaison Agency; 96 Roger Ressmeyer/ Corbis; 97 HRW Photo; 98(c), Andy Christiansen/HRW Photo; 98(l), John Blaustein/Liaison Agency; 98(r), Mark Lewis/Stone; 99(tc), James Randklev/Stone; 99(r), Bruce Hands/Stone; 99(br), John Zoiner Photographer; 99(bl), Ed Malles/Liaison Agency; 100(cl), Andy Christiansen/HRW Photo; 101 Telegraph Colour Library/FPG International; 102(cr, bl), John Zoiner; 104(t), Horst Schafer/Peter Arnold, Inc.; 104(ct), Paolo Koch/Photo Researchers, Inc.; 104(cb), Brian Parker/Tom Stack & Associates; 104(b), C. Kuhn/Image Bank; 105(bl), Mark A. Leman/Stone; 105(br), Tim Eagan/Woodfin Camp & Associates; 106(tr), Adam Hart-Davis/Science Photo Library/Photo Researchers, Inc.; 106(bl), James Stanfield/National Geographic Image Collection; 107(tc, tr), Victoria Smith/ HRW Photo; 108(bl), Sylvain Coffie/Stone; 109(tr), Tom Myers/ Photo Researchers, Inc.; 110(c), Alex Bartel/Science Photo Library/Photo Researchers, Inc.; 110(bl), Joyce Photographics/Photo Researchers,Inc.; 111(bl), Hank Morgan/Science Source/Photo Researchers, Inc.; 112, Mark Lewis/Liason International; 113(tr), Craig Sands/National Geographic Image Collection; 113(cl), Tom Bean; 114(cr), G.R. Roberts Photo Library; 116 Richard Hutchings/HRW Photo; 120(tl), John Blaustein/Liaison Agency; 120(tr), Tom Myers/Photo Researchers, Inc.; 122(tr), SuperStock; 122(c), Bedford Recycled Plastic Timbers; 122(b), Kay Park-Rec Co; 123 Culver Pictures, Inc.

Labook: "LabBook Header": "L", Corbis, "a", Letraset-Phototone, "b" and "B", HRW, "o" and "k", Images ©2001 PhotoDisc/HRW, Inc. 125(cl), Michelle Bridwell/HRW Photo; 125(br), Image ©2001 Photodisc, Inc.; 126(bl), Stephanie Morris/HRW Photo; 127(tr), Jana Birchum/HRW Photo; 135(tr), Tom Bean/DRK Photo; 135(br), Darrell Gulin/DRK Photo, 136(br), Mark Heine.

Appendix: p. 158 CENCO

Sam Dudgeon/HRW Photos: all Systems of the Body background photos, p. viii-1, 80(tr), 87(br), 107(cr, c), 109(b), 114(tl), 117, 124, 125(bc), 126(br, tl, tr), 127(tl), 128(bc, tr), 129(tr, br), 130, 131(tr, br), 137, 141(br)

Peter Van Steen/HRW Photos: p. 38, 81(bl, bc, br), 84(tl), 89, 91, 127(b), 133, 134, 141(tr)

John Langford/HRW Photos: p. 125(tr)

Acknowledgements continued from page iv.

Alyson Mike
Science Teacher
East Valley Middle School
East Helena, Montana

Donna Norwood
Science Teacher and Dept. Chair
Monroe Middle School
Charlotte, North Carolina

James B. Pulley
Former Science Teacher
Liberty High School
Liberty, Missouri

Terry J. Rakes
Science Teacher
Elmwood Junior High School
Rogers, Arkansas

Elizabeth Rustad
Science Teacher
Crane Middle School
Yuma, Arizona

Debra A. Sampson
Science Teacher
Booker T. Washington Middle School
Elgin, Texas

Charles Schindler
Curriculum Advisor
San Bernardino City Unified Schools
San Bernadino, California

Bert J. Sherwood
Science Teacher
Socorro Middle School
El Paso, Texas

Patricia McFarlane Soto
Science Teacher and Dept. Chair
G. W. Carver Middle School
Miami, Florida

David M. Sparks
Science Teacher
Redwater Junior High School
Redwater, Texas

Elizabeth Truax
Science Teacher
Lewiston-Porter Central School
Lewiston, New York

Ivora Washington
Science Teacher and Dept. Chair
Hyattsville Middle School
Washington, D.C.

Elsie N. Waynes
Science Teacher and Dept. Chair
R. H. Terrell Junior High School
Washington, D.C.

Nancy Wesorick
Science and Math Teacher
Sunset Middle School
Longmont, Colorado

Alexis S. Wright
Middle School Science Coordinator
Rye Country Day School
Rye, New York

John Zambo
Science Teacher
E. Ustach Middle School
Modesto, California

Gordon Zibelman
Science Teacher
Drexel Hill Middle School
Drexell Hill, Pennsylvania

Self-Check Answers

Chapter 1—Interactions of Living Things

Page 9: Humans are omnivores. An omnivore eats both plants and animals. Humans can eat meat and vegetables as well as animal products, such as milk and eggs, and plant products such as grains and fruit.

Page 10: A food chain shows how energy moves in one direction from one organism to the next. A food web shows that there are many energy pathways between organisms.

Page 15: 1. If an area has enough water to support 10 organisms, any additional organisms will cause some to go without water and move away, or die. 2. Weather favorable for growing the food that deer eat will allow the forest to support more deer.

Chapter 2—Cycles in Nature

Page 36: The main difference between primary and secondary succession is that primary succession begins with the formation of soil. Secondary succession begins on preexisting soil, such as when an existing community is disrupted by a natural disaster or by farming. Pioneer species in primary succession are usually lichens, which begin the formation of soil. Pioneer species in secondary succession are usually seed plants, which germinate and take root in the soil.

Chapter 3—The Earth's Ecosystems

Page 53: Deciduous forests tend to exist in mid-latitude or temperate regions, while coniferous forests tend to exist in higher, colder latitudes, closer to the poles.

Page 58: Answers include: the amount of sunlight penetrating the water, its distance from land, the depth of the water, the salinity of the water, and the water's temperature. 2. There are several possible answers. Some organisms are adapted for catching prey at great depths; some feed on dead plankton and larger organisms that filter down from above; and some, such as the bacteria around thermal vents, make food from chemicals in the water.

Chapter 4—Environmental Problems and Solutions

Page 77: 1. We use nonrenewable resources when we burn fossil fuels when driving or riding in a car or burning coal for heat. When we use minerals that are mined, we are using a nonrenewable resource. Pumping ground water is another use of a nonrenewable resource, if the water is used faster than it is replenished. 2. If a nonrenewable resource is used up, we can no longer rely on that resource. Certain oil and coal deposits have been building since life began on the planet. It may take hundreds of years to replace a mature forest that can be cut in a day.

Page 83: 1. Turn off lights, CD players, radios, and computers when leaving a room. Set thermostats a little lower in the winter (wear sweaters). Don't leave the refrigerator door open while deciding what you want. 2. plastic bags, rechargeable batteries, water, clothing, toys; The difference between a reused and a recycled object is that a reused article may be cleaned but is basically unchanged. A recycled article has been broken down and re-formed into another usable product.

Chapter 5—Energy Resources

Page 113: Both devices harness energy from falling water.